A JUMBLE OF NEEDS

A Jumble of Needs

· · · ·

Women's Activism and Neoliberalism in the Colonias of the Southwest

Rebecca Dolhinow

University of Minnesota Press
Minneapolis
London

Portions of chapter 2 were previously published as "Women's Leadership in New Mexican Colonias," in *Women and Change on the U.S.–Mexico Border: Mobility, Labor, and Activism*, ed. Doreen Mattingly and Ellen Hansen (Tucson: University of Arizona Press, 2006); copyright 2006 by the Arizona Board of Regents; reprinted by permission of the University of Arizona Press. Portions of chapter 4 were previously published as "Caught in the Middle: The State, NGOs, and the Limits to Grassroots Activism along the U.S.–Mexico Border," *Antipode* 37, no. 3 (2005): 558–80; reprinted by permission of Blackwell Publishing.

Published by the University of Minnesota Press
111 Third Avenue South, Suite 290
Minneapolis, MN 55401-2520
http://www.upress.umn.edu

Library of Congress Cataloging-in-Publication Data

Dolhinow, Rebecca.
A jumble of needs : women's activism and neoliberalism in the colonias of the Southwest / Rebecca Dolhinow.
    p. cm.
Includes bibliographical references and index.
ISBN 978-0-8166-5057-6 (hc : alk. paper) — ISBN 978-0-8166-5058-3 (pb : alk. paper)
1. Women in community organization—New Mexico—Case studies. 2. Community leadership—New Mexico—Case studies. 3. Women civic leaders—New Mexico—Case studies. 4. Working poor—Housing—New Mexico—Case studies. 5. Mexicans—Social conditions—New Mexico—Case studies. 6. Slums—Social aspects—New Mexico—Case studies. I. Title.

HQ1240.5.U6D65 2010
305.48'9694209789-dc22   2009016361

Printed in the United States of America on acid-free paper

The University of Minnesota is an equal-opportunity educator and employer.

16 15 14 13 12 11 10   10 9 8 7 6 5 4 3 2 1

In memory of Carlos: your spirit, experience, and wisdom touched so many in the colonias and in them your work lives on.

Contents

Acknowledgments

Many people have shaped this book, and they all have my sincere appreciation. But without the women who organize and bring attention to their lives in the colonias, there would be no book. The women whose lives are discussed here were much more than just my informants: they are my friends and women from whom I learned some powerful life lessons. This book is not only about them—it is for them. It is also for my friends at the Community-Organizing Group, who do great work under difficult circumstances; I hope this book is of value to them. I wrote this book for those who live and work in the colonias.

I began the journey that ends in this book while in graduate school, and I must thank my chairs for their encouragement and guidance. Gill Hart and Allan Pred shaped my thinking as a geographer in important and vital ways. I wish Allan were with us to see the final result. I sincerely thank Caren Kaplan, a splendid mentor, who guides with sincere wisdom and understanding. Over many cups of chai (expertly prepared by her partner Eric), Caren helped me turn a pile of pages into a manuscript. While I was at the University of California at Berkeley, many friends, colleagues, and others helped me, and a few must be thanked: Aaron Bobrow-Strain, Jessica Teish, Susan Shepler, Ruthie Gilmore, Wendy Wolford, Heather Merrill, Delores Dillard, Nat Vonnegut, Carol Page, Caroline Altman, Sarah Baughn and Eric Armstrong, Brenda Milis, Charlie and R. J. Harrison, the Morris family, Pete and Ann Haley, Elsworth Ray, Miriam Joscelyn, and Kristin Ghodsee. At Berkeley I met my writing partner, Nitasha Sharma. No one else knows this project as well as she does. No one else has read this book as many times as she has! Thank you, Tasha. Those I have not thanked directly know who you are and know you are also appreciated with all my heart.

During my fieldwork in New Mexico, several people outside the colonias supported me in my work and became great friends. My experiences in Las Cruces would not have been the same without the Perez family, Estella's red chile, and Ernesto's sense of humor. Sarahh Scher was there when I really needed a girlfriend, and she introduced me to *Farscape*.

I have since left Berkeley for Southern California and now have a new group of friends who keeps me on track: Jessica and Jeanine, Natasha, Scott, Ashton, Maddox and Ethan, Jessica, Mihir, Bella, and Miki Zoe and Mary and Tweedy and Grace. Finally, I thank my inspiring colleagues, Renae Bredin, Marjorie Jolles, Donna Nicol, and Karyl Ketchum; my terrific students; and my research assistant Maribel Reyes in the women's studies program at Fullerton. My students help me to see the worlds of activism and women's studies in new ways that illuminate both this project and my life. Thank you—and keep it up! Of course, I must also thank the women at Fullerton who have made life so much easier for me: Judy, Joan, Sharon, and Gloria. Janice helped me to keep true to myself and this project, often when I was not sure I could, and for this I am grateful. In the end, what you see here is all my own, and I take responsibility for any errors or oversights.

This research was made possible by funding from the National Science Foundation, the University of California Institute for Labor and Employment, the University of California Mexus Program, and the Center for Latino Policy Research at the University of California at Berkeley. At the California State University at Fullerton, I was able to work more productively due to release time generously provided by both the Office of the President and the Dean's Office in the College of Humanities and Social Sciences. I am grateful for all of this financial support.

The arguments in this book evolved tremendously during my year as a Rockefeller fellow at the University of Texas at Austin. While working with the other scholars in the "Race, Rights, and Resources in the Americas" program, I could think seriously about what it means to be involved in research based in social justice with other academics who are deeply committed to similar work. During my time in Texas, I was privileged to work with Shannon Speed, Angela Stuesse, Charlie Hale, Virginia Raymond, Laura Padilla, Olga Herrera, Jim Lee, Julie Cho, and David Kamper.

I thank Jim Lee for introducing me to Richard Morrison, and I thank Richard for walking me through the first steps of my first book. I had the pleasure of working with Jason Weidemann at the University of Minnesota Press, and I thank him for all of his guidance and support. Jason and Richard found such appropriate and valuable readers for this manuscript;

one reader in particular offered the most insightful commentary, and I thank her for her generous guidance.

The final revisions to the manuscript would have been much more difficult and time consuming were it not for a summer scholarship at the School for Advanced Research in Santa Fe, New Mexico. The time, space, resources, and amazing cohort of colleagues the School for Advanced Research offered were invaluable to the completion of the manuscript and my peace of mind. There is nothing like literally having a room of one's own in which to write, especially when it comes with a bird feeder and dozens of colorful birds right outside the window.

None of this would have been possible without the support of my parents, who introduced me, at a very early age, to the academy. My father's interest in politics, science, philosophy, and much more and the never-ending stack of books on the coffee table shaped my interests today. In many ways my mother is my greatest academic role model: she shows me that compassion and rigor are both important to an academic career. Thank you.

Abbreviations

Pseudonyms are used in place of common abbreviations for some organizations in order to maintain anonymity.

AHI	Affordable Housing, Incorporated (pseudonym)
BIP	Border Industrialization Program
BLM	Bureau of Land Management
BWG	Border Water Group (pseudonym)
COG	Community-Organizing Group (pseudonym)
GIS	Geographic Information Systems
ICE	U.S. Immigration and Customs Enforcement
IGF	Independent Gay Forum
INS	U.S. Immigration and Naturalization Service
IRCA	Immigration and Reform Control Act
NAFTA	North American Free Trade Agreement
NGO	nongovernmental organization
NOS	National Origins Act

Introduction

Cuando en el río va con el corriente.
When in the river, go with the current.

"I HOPE YOU'LL DESCRIBE HOW the chile feels when it starts to rot!" Estella was clearly excited about the prospect that I might write about our experiences together over the last year, and she had very specific ideas about what I should document. Estella usually had strong ideas about everything; this was something I learned about her very quickly. Estella was not a particularly large woman. She was a little below average height and a bit stocky, but her personality made her seem larger than life and gave her a very commanding presence. For Estella, the daily lived practices, the taken-for-granted, the textures, smells, and sounds of her colonia needed to be documented alongside her work as an activist. So it was crucial that my depiction of her at-home *ristra* production business include not only the economic and social aspects of this culturally and geographically specific form of income supplementation but also a detailed description of the milieu. This would have to include the texture of the chiles after a few days in the sun, like *gusanos*, or worms; squishy, slippery, and not too nice smelling either—but something I had to feel for myself, with both hands, to really know when they were too far gone to be useable. At least that is what Estella said as she pushed my hands into the rotten chile, smiling and laughing loudly all the while.

In the late fall Estella, a central leader and activist in the colonia of Recuerdos, turned the front porch of her father's trailer into a lively production line for *ristras*. Ristras are red-chile wall hangings that can be both decorative and used in cooking and are ubiquitous signs of the intersection of New Mexico's agricultural and tourist economies. In tourist towns throughout the Southwestern United States, ristras are a common site that fetch big bucks, particularly when they are as fresh and bright red as they are in the fall. Production begins with a call to a local chile producer, who will drop off an enormous crate of red chile, which then must be sorted by size, a job at which the children excel with a little guidance. Next the ristras are constructed using string and brute force through an artful weaving job that takes more practice than it would seem. Finally the chile producer, also known as the middleman, returns to buy back the ristras for a few dollars each. He will go sell them in the north of the state in tourist destinations such as Santa Fe and Taos for twenty dollars or more, making a handsome profit. The women will have made a few dollars an hour and had a lot of time for *chisme* (gossip) while also keeping up with child care and household chores. Although ristras would never make Estella or the women who joined her rich, there were several reasons why they made sense as a seasonal project. Estella and her neighbors were primarily stay-at-home mothers and wives, whose husbands were field workers, and, like everyone in Recuerdos and nearly everyone in all of New Mexico's colonias, they were also Mexican immigrants. Immigrant field labor families make very little money, so, when opportunities such as ristra making appear, women often take advantage of them. Through their ristra business, women like Estella are creating a "productive" niche for themselves in an agricultural economy, which is heavily dependent on the labor of their husbands, fathers, and sons. These women are already indispensable to this economy, as they are at the base of the social reproduction that keeps the flow of cheap Mexican labor available. This is the heart of colonia communities: the social reproduction of agricultural labor.

Ristra making is a particularly good employment match for a busy mother because she can make a couple of ristras, which take about fifteen to twenty minutes to finish, and then take a break and change a diaper, do a quick load of laundry, or put on a pot of soup. While I was working on the ristra production line, there were always small children around playing together; it was a makeshift daycare, with women coming and going as they saw fit. In their ristra production business, the women created an entrepreneurial space so they could take advantage of their role as cheap labor and use it to suit their needs as mothers and caretakers. This was also valuable to Estella as an activist and community leader because it allowed her easy access to her community. She could keep in touch with the goings-on in the community and keep up with the needs of her neighbors. Estella knew that, in the colonias, change was directly related to the needs and desires of women because the women were the movers and shakers in the colonias. Women like Estella are leaders in colonias across southern New Mexico, organizing their communities to create the necessary infrastructure for healthy living.

The federal Office of Housing and Urban Development's (HUD) definition of a colonia is an unincorporated community within 150 miles of the U.S.–Mexico border that lacks one or more of the following: potable water, waste water, paved roads, and safe and sanitary housing.[1] This is the definition used by HUD to determine colonia status and eligibility for federal monies for improvements.

This definition, like the colonias of the U.S.–Mexico borderlands in general, is characterized by lack and absence. If all you knew about the colonias was this federal definition, and you were to close your eyes and try to imagine a colonia, what would you see? You must remember, you would not have been told anything about why these communities lacking infrastructures exist, or who developed them, so you would not know colonias are inhabited by Mexican immigrants or that the residents are poor farm workers. You might be able to muster up an image of a rather rustic kind of community with dirt roads, probably in a desert-borderlands setting with rundown houses and wells for water and outhouses. But after that, there are holes in the image because of a lack of information. The 1990 U.S. General Accounting Office report often is cited as the first federal acknowledgment that colonias were a growing problem in Border States. It states, "Available data, although limited, indicate that the residents of colonias are mostly Mexican-American, many work as seasonal farm laborers, and many have incomes below the poverty line" (U.S. General Accounting Office 1990).[2] In Texas alone, four hundred thousand people live in colonias.[3] In New Mexico, where this research took place, the colonia population tops seventy thousand.[4]

This book is the story of what happens when too much is missing from the life of a neighborhood, of so many neighborhoods, and of a surplus of people (Gilmore 2007). The story that will unfold is of the abandonment of a group of people, Mexican immigrants, by the state, which is deeply embedded within the neoliberal political project. The story then follows the actions of the same state that allowed, and even encouraged, the private sector to create substandard housing for this group of essential workers and then partnered with a third sector of nongovernmental organizations (NGOs) and community leaders to provide for these workers the rest of the services they require in order to survive.

This story is told through an ethnography, the focus of which is the production of a place, a colonia, as a site of convergence for the processes of neoliberal governance, transnational flows of capital and labor, and the role of women leaders in these processes on the U.S.–Mexico border. I argue that these processes of governance in which women leaders take part limit the development of politically conscious, community-level progressive social change that confronts systems of power. Through their daily interactions with NGOs, who have, often unwittingly, become partners in systems of neoliberal governance, women leaders, too, become conduits for such forms of governmentality.

Colonia-like communities have existed along both sides of the border since the early 1900s. In their current and most popular form, colonias are a mid- to late twentieth-century phenomenon in Texas and New Mexico. Despite this long history, they are the topic of very little scholarly work. Given the large colonia population on the border and the increasing importance of political and economic relations between the United States and Mexico, the lives of Mexican immigrants in colonia communities require attention. Just days before the events of September 11, then-President of Mexico Vincente Fox spoke to a joint session of the U.S. Congress, a quite remarkable and rare event. But it made perfect sense that it should be the president of one of our nearest neighbors and our second-largest trading partner at the time that would merit such unusual treatment. President Fox spoke of the growing economic and political ties between the two countries that share a two thousand–mile border. After his election in 2000, President George W. Bush[5] (the former governor of Texas) and President Fox met several times and discussed the future of U.S.–Mexico relations. Fox counted on Bush's experience and support when he proposed a new Guest Worker Program between the United States and Mexico and amnesty for undocumented Mexican migrants in the United States (Felch 2002). The speech was well received and the future of U.S.–Mexico relations looked brighter than it had in years. Then everything changed on September 11, and it was not until immigration reforms began again in earnest in 2007 that U.S.–Mexico relations and the place of Mexican immigrants in matters other than "homeland security" became a matter of national debate once again.

Discourses surrounding immigration may change with time, and the focus of xenophobic outrage among racist and violent anti-immigrant groups may vary as American foreign policy shifts, but one thing is certain: Latinos in the Southwest have been steadily growing in numbers for decades. While the rest of the nation is just waking up to its love–hate relationship with immigrants, the Southwest has been aware of this relationship for years, and the colonias developed in part as a solution to this problem of increasing xenophobia. They keep this growing immigrant population out of sight. Colonias are often built in out-of-the-way and unseen places, behind agricultural fields, and in little valleys. In this way, they are out of sight and out of mind. Often, what appears to be a group of deserted trailers in the midst of the dry desert brush is actually a colonia.

The majority of New Mexico's colonias and those in this book exist in Doña Ana County, which, according to the U.S. census, is 3,815 square miles.[6] All three colonias in this study are in the Mesilla Valley, an agricultural area created by the flood plain of the Rio Grande as it runs through the center of the county. This valley supports a range of agriculture, from the green chile for which the north of the valley is famous and the residents of Recuerdos pick annually, to the pecans grown in the "world's largest pecan orchard" in Stahmann's Farms in the south, where residents from Los Montes and Valle de Vacas have worked over the years. All of this agriculture is irrigated to some extent by the Rio Grande, making it an obvious choice for developers to locate colonias near the fields, which are in turn near enough to the river for gravity irrigation systems to function. Colonias thus often develop along the paths of rivers or by irrigation ditches.

In 2000, when this research was conducted, the total population of Doña Ana County was 174,682.[7] Of this total, 118,478 were Caucasian; 2,723 black/African American; 2,580 American Indian or Alaska Native; 1,330 Asian; and 117 identified as Native Hawaiian/Pacific Islander. This left 110,665 Latinos in the county, a percentage nearly equal to that of whites.[8] In the same year in Doña Ana County, the percent of foreign-born residents was 18.7 percent, and, in New Mexico as a whole, it was only 8.2 percent. In 2006, Doña Ana County was 65 percent Latino and New Mexico was 44 percent.[9] Based on these numbers, it would be safe to say that the majority of the disparity in foreign-born populations in Doña Ana County was in the Latino population. In Doña Ana County, 45 percent of the Latinos are Mexican, according to the 2000 census data, and this fits with my experiences in the colonias, which are almost wholly Mexican communities.[10] On the border, in colonias where poverty is an endemic problem and agriculture jobs are a mainstay, a high school education can seem like a waste of time, and this is reflected in the statistics. The percent of the population of Doña Ana County over the age of twenty-five with a high school diploma is 70 percent, which is 8.9 percent below the state average. The percent below the poverty line in 2004 was 16.7 percent for New Mexico as a whole, and 23 percent for Doña Ana County.[11]

Many of the poverty-stricken in Doña Ana County live in colonias. There are thirty-four designated colonias in Doña Ana County. This research took place in three, all of which are within fifty miles of Las Cruces, the third-largest city in New Mexico and the largest in Doña Ana County. Las Cruces makes a triangle of sorts with El Paso, Texas, and Ciudad Juarez, Mexico, to create one of the border's busiest and most sprawling metropolitan centers. It is against this backdrop that the region's colonias developed. For the residents of Recuerdos, Los Montes, and Valle de Vacas, the El Paso–Juarez crossing was the closest and easiest way into Mexico. In 2000, when this research took place, crossing the border was still a rather routine and simple event for a person with documents. Shopping, dining, and visiting the dentist were all valid reasons for a trip *al otro lado*, to the other side. Since September 11, these trips are no longer routine events and require a good deal of planning. Simply shopping is not reason enough to warrant the lines and harassment experienced in the border crossing process. The majority of colonia residents are documented, yet they still encounter resistance at border crossings and at checkpoints in the United States. I was often asked if the women with whom I worked were documented, and the answer was all of them but one. This is an important question on many levels, as we will see. Although most colonia residents were documented, as with any immigrant community, there were relatives and friends that came through the colonias whose documentation status was irregular. For this reason, colonia residents were vigilant about issues pertaining to immigration and the Border Patrol in particular. Colonias are communities of Mexican immigrants living and working in the United States, communities of immigrants living on land they are buying to pass down to their children. Procuring documentation and regularizing their status is an important first step in this process for many colonia residents.

The Project

Through the detailed study of daily practices, my research examines the moment in which a particular set of processes and geography meet to create colonias. In particular, the processes of global economic restructuring are crucial to understanding how these communities came to exist in the first place. The power of neoliberal policy can be felt every day in the resource deprivation that characterizes these settlements. It is not possible to understand the complexities of a situation by the simple identification of large-scale processes at work. Rather, it is necessary to acknowledge the multiple ways in which people engage with and shape these processes daily. Transnational processes unfold differently in every locality. Allan Pred acknowledged the complexity of the production of the local and its relations to nonlocal processes in his work in Sweden: "What sets the hypermodern present apart is the intense intermeshing of the 'local,' the subnational, the national, and the 'global'—is the variety, spatial extensiveness, complexity, and instability of the interactions and interrelations through which the 'local' is constituted and reproduced" (2000, 23). In the colonias, this present is clearly the result of many transnational and global processes reworked by and through local processes and institutions.

At each particular historical, geographical, cultural, or economic moment, structures and processes are materially and concretely manifest. Abstract theories cannot reveal these intricately localized manifestations; only the study of daily practices can accomplish this. The many combinations of actors, power relations, histories, and geographies that make each moment and place specific lie at the core of my project.

The central challenge of this study is twofold: First, I examine the construction and gendering of leadership through the constant reworking of neoliberal restructuring, transnational connections, and NGO interventions within locally interpreted ethnic and gendered relationships. Second, I question the effects of these relationships and interventions on the activism of leaders in the colonias.

The focus on the particular time and place at which colonias developed accomplishes two goals. First, an examination of the specific, multiple-scaled processes involved in colonia creation sheds light on how the colonias developed as an affordable housing alternative for the region's population of Mexican working poor and as self-sufficient communities that serve the interests of the neoliberal political project. Second, this focus acknowledges the connections between the processes that produce colonias as physically isolated communities and colonia leaders as political subjects who unintentionally support and reinforce neoliberal policy. The key processes implicated here are global economic restructuring, the migration we see in its wake, the growth of neoliberal policy from the early 1980s, the attending expansion of the role of NGOs in many communities, and neoliberal governance as a key element of the neoliberal political project. Under these specific conditions, NGOs, the state as it supports the neoliberal political project, and colonia leaders come together to limit development of progressive social change that confronts systems of power at the community level in the colonias.

The focus on women's activism and the role of NGOs in this activism is an important intervention in part because it situates the literature on women's activism in the United States within debates in the literature on women's activism in the global south, particularly in relation to development-oriented NGOs. The relationship that I found between women leaders and the NGOs that served these communities shared much in common with their sister organizations in development programs in countries across Africa and South Asia (Mohan 2002; Rankin 2004).

This research contributes to the literature on the transnational nature of globalization on the border in order to clarify how women-led immigrant communities fit into larger scale processes of global economic restructuring. Closely tied to this literature is a body of scholarship that focuses on the ties between globalization, neoliberal policy, and the neoliberal political project, including discussions of the relationships between locally based NGOs and broad-scale neoliberal agendas. To this discussion I hope to contribute a detailed feminist examination of the role that NGOs play in the organizing of colonia communities and the position they inhabit as a mediator between women colonia leaders and the state. A study of daily life in the colonias brings the literature on governmentality and neoliberalism into conversation with an examination of the conditions that produce women as leaders in these sites. This is an ethnography of the production of a place, a colonia, as a site of convergence for the processes of neoliberal governance, the formation and actions of an institution (an NGO), and the role of women as leaders in the era of flexible globalization on the U.S.–Mexico boundary.

The Processes at Work

Colonias exist both because, and at the mercy, of the processes of globalization along the U.S.–Mexico boundary. Colonia leaders organize their communities to gain the resources they lack due to the effects of neoliberalism. In the colonias we see the effects of an intricate web of processes led by neoliberal cutbacks that seriously affect women's daily situated practices. Yet by organizing their colonias to provide what they require for themselves, the actions of colonia leaders appear to be letting the state off the hook.

In his discussion of Marx's 1857 Introduction to the Grundrisse, Stuart Hall focuses on the necessity and importance of complexity in the analysis of "production": "Essentially, we must 'think' the relations between the different processes of material production as 'members of a totality, distinctions within a unity'. That is as a complexly structured differentiated totality, in which distinctions are not obliterated but preserved—the unity of its 'necessary complexity' precisely *requiring* this differentiation" (1974, 146, emphasis in the original). In the same way that Hall points to Marx's awareness of the complexity of the processes of material production, I stress the necessity of viewing the many processes involved in daily life in the colonias as "a complexly structured differentiated totality." According to Hall, Marx conceptualizes the links between structures as a "complex unity": "For Marx, two different terms or relations or movements or circuits remain specific and different: yet they form a 'complex unity'" (146).

The relationship between neoliberalism and NGOs is a good example of this "complex unity." Without neoliberal cutbacks, NGOs would not be as in demand to provide social services. Without NGOs, it would be more difficult for the state, as it serves the neoliberal political project, to cut back its social service provision. Yet it is important that NGOs and neoliberal policy at least appear distinct from each other in order for the state to create the ideal of the self-sufficient citizen that characterizes the neoliberal political project. Hall links this method to material practices: "This means that, in the examination of any phenomenon or relation, we must comprehend *both* its internal structure—what it is in its differentiatedness—as well as those other structures to which it is coupled and with which it forms some more inclusive totality. Both the specificities and the connections—the complex unities of structures—have to be demonstrated by the concrete analysis of concrete relations and conjunctions" (147).

My research reflects Hall's emphasis on the materiality of Marx's methods and supports the claims that phenomena must be examined alone (in its differentiated form) and as it is connected to other phenomena. With regard to the colonias, NGOs and the state often described individual community projects as "empowering," "politicizing," or "uniting." But my research points to a much less optimistic picture that only becomes clear when individual projects and colonias are examined as part of a greater whole. This whole encompasses colonias and other marginalized communities across the globe. Without the necessary attention to the "connections," a very misleading picture can arise.

Globalization

While globalization studies often exemplify the worst of overgeneraliza-
tion, more specific examples of globalization can demonstrate the unify-
ing work that Hall discusses. When conceptualized as a complex set of
multiple-scaled processes, globalization can be used to help theorize the
interactions of disparate communities across the world. This use can allow
for similarities and differences to be recognized and creates the potential
for coalitions working for change and social justice (Katz, 2001a). In the
colonias, the economic restructuring and neoliberal policies associated
with globalization factor strongly into the creation of both these informal
settlements and the development of women's leadership.

Particularly important for my work are feminist critiques that question
both less-nuanced uses of globalization and the way in which globalization theory is often used without attention to gender. According to Richa
Nagar et al., globalization theory often constructs certain places, usually
in the global south, as "marginal . . . and as passive recipients of, or as
irrelevant to, globalization" (2002, 265). These passive areas are depicted
as simply accepting globalization, its benefits, and rewards alongside its
difficulties and disadvantages. They are seen as incapable of affecting
change on global processes that exist in a somehow larger-than-life realm.
This spatial marginalization works hand in hand with a gender bias that
"sidelines" women and gender analysis more generally, in this way creating a "double marginalization" (265). Alongside this critique, Nagar et al.
highlight "the complexities and contradictions that are inherent in actors'
relationships to globalization" (270). The goal of their focus on these complexities is to "highlight the ways in which global processes can be liberating as well as exploitative" (270).

An acknowledgment of the complexities and contradictions of global
processes is crucial. A feminist analysis of globalization can provide this
detailed understanding: "A feminist analysis therefore fundamentally
changes the nature of understandings of economic globalization. Above
all, it entails a shift from a straightforward, linear, master narrative toward
diverse and possibly conflicting accounts" (280). These authors hold that
feminist accounts of globalization are necessary because their focus on
situated practices tends to highlight the nature of power relations, and
this is important as we look to question existing understandings of globalization and the neoliberal political project. As feminist approaches

to globalization studies open up conflicting accounts, they also open up investigations into whose accounts do and do not "count," and it is questions such as these that bring to the forefront larger questions about the production of power and knowledge as political processes that are central to understanding how neoliberal globalization affects all people (Nagar 2002). The acknowledgment of the necessity of these "possibly conflicting accounts" is especially important in the research presented here, where the goals of NGOs and grassroots activists, like rivers in relation to their beds, meander back and forth, sometimes closer to one side and sometimes closer to the other.

The feminist inquiry Nagar et al. call for is central to my work in the colonias, where the relationships between gender and globalization are played out in the daily lives of women leaders. Men and women interact in different ways with the processes of global economic restructuring. In the colonias, just as in many marginalized communities of color in the developed and developing world, women are responsible for the day-to-day basics of running a household in a period of economic instability and high unemployment rates. As global economic restructuring interacts with gendered divisions of labor in the home and community, often the result is more work for women along with fewer resources to work with (Christopherson 2002). Many of the debates on the gendered nature of globalization revolve around women's central role in the processes of social reproduction. Cindi Katz points to the differing costs of social reproduction as central to the movement of capital along the paths of globalization (2002). According to Katz, in an effort to cut costs, capital will often move to the place where social reproduction is cheapest. The costs of social reproduction are of central importance and must be controlled and reduced through the movement of capital. It is women and children whose lives most concretely intersect with the movement of capital as it searches out the cheapest places to reproduce the necessary labor: "In fact, it is when factoring in the costs of social reproduction that the expendability of this population comes directly to light" (253). Although Katz's work focuses on children, the same arguments can just as easily be applied to women activists in any working-poor community.

Seemingly "placeless" capital no longer sees the social reproduction of the working poor as its responsibility. When capital becomes more mobile, as we see in the current phase of globalization, the social reproduction of any one particular group becomes less important. Melissa

Wright's insightful work on women working in global factories makes this abundantly clear as women working in both Mexico and China experience a lack of resources and literally become "disposable" within the processes of global production (2006).

Neoliberalism

Politics and economics move hand in hand in globalization, often through discourses of "development" and the state's role in the purposefully controlled movement of capital in the processes of social reproduction. In discussing neoliberalism, perhaps the most important process at work in the colonias, it is important to acknowledge that it, too, has more than economic and political aspects (Harvey 2005). As an economic project, its primary goal is the redesign of capitalism on an international scale as a free-market enterprise. On the political side this goal is attained through similar means, but the end is the movement of power to the wealthy, that is, the truly economic elite. A second, but not secondary, goal, particularly in relation to this research on community organizing and activism, is the work of neoliberalism as a political project to destroy all collectives. In a 1998 article in *Le Monde*, Pierre Bourdieu described neoliberalism as a "political project," a "scientific programme, converted into a plan of political action, an immense *political project*" (1998, 2). This project, Bourdieu explained, calls "*into question any and all collective structures* that could serve as an obstacle to the logic of the pure market." This is only one of many important aspects of the neoliberal hegemony that we see at work in the functioning of the state, the workings of the economy, and in many social and cultural institutions (2). In the colonias, this political project (with its focus on the individual and the desire to destroy the collective) means abandonment. As communities of poor agricultural workers, neoliberalism has little obvious use for colonias and even less desire to waste money or resources on them. In the neoliberal system, they are left to fend for themselves. Yet, on closer examination, these resource-lacking settlements fit perfectly into the basic tenets of neoliberal politics.

What Is Neoliberalism?

Neoliberalism takes its name and direction from liberal economics as theorized by Adam Smith in the late eighteenth century. Liberal economics, and, in turn, neoliberal economics, are founded on the premise that regulation of markets is detrimental to an economy and that free, or unregulated, markets are most favorable. The markets in question are those of the capitalist system. From this premise, the theory expands to include a preference for limited government involvement in citizens' lives in order to keep government small and out of service provision, which liberal and neoliberal theory holds ought to be privatized, or "rolled back," in order to function in the most efficient manner possible (Peck 2004). As neoliberalism moves the ideas of liberal economics into the twentieth and twenty-first centuries, it also, according to Larner, creates "forms of political-economic governance premised on the extension of market relationships" (2000, 5). Neoliberalism is a hegemonic political project that encompasses economic, political, and governmental systems. In the colonias, we see all three aspects of neoliberalism at work, but it will be neoliberal governance as it is manifest in their daily practices that disables the work of leaders in the most obvious ways.

Moving from the eighteenth to the twenty-first century we see the ideas of liberal economics, and its attending ideologies of politics and governance, move through popularity in the United States in waves. Liberal economic ideologies dominated in the United States up until the 1930s during the Great Depression, which opened a window for economic and political regime change. Growing inequality led to concerns that a new system might be necessary, one that could redistribute wealth in a more equitable way. The system that was developed did this through state intervention, the antithesis of liberal economics. This system was based on the principles of the English economist John Maynard Keynes and his American disciple John Kenneth Galbraith. During the Keynesian era in the United States, both economic and political interventions were made to promote market growth and fight the unemployment and deflation of the Depression. Keynes postulated that markets would only grow if people were actually spending enough money, and that during the Depression people were not doing this. If people were saving too much money, or were not being paid enough to spend money freely, there would not be enough money moving through the market and the market would suffer for it. In order to stop this cycle of minimal demand, Keynes suggested state intervention in wages through programs like social security, unemployment, and employment training, all of which created more disposable income and, in turn, more demand. These programs also created what would become known as a "social safety net," which helped citizens when they were down on their luck. Keynesian economics was a system based on the ideas that the state should help the people to help themselves and in turn create a greater good, a collective ideal. After World War II and the Bretton Woods Conference where Keynes's ideas were used to design the plans for the restoration of Europe, Keynesian economics went global (Conway and Heynen 2006; Peet 2003).

Yet, by the 1960s, the end of the Keynesian era was already in sight. By the 1970s, growth in the United States had severely slowed down and was nearing collapse with inflation and unemployment on the rise while interest rates and profits were falling. The upper classes had watched their profits diminish and their fortunes be "redistributed" for long enough. In his book, *A Brief History of Neoliberalism*, David Harvey, like Bourdieu, argues that neoliberalism should be viewed as "a *political* project to re-establish the conditions for capital accumulation and to restore the power of economic elites" (2005, 19, emphasis in the original). If we view neoliberalism in this way, then 1979 was a watershed year. In May of that year, Margaret Thatcher was elected Prime Minister in Britain, and hers was a clearly neoliberal and anti-Keynesian agenda. One year later in the United States, Ronald Reagan beat Jimmy Carter for the presidency, and thus began the neoliberal conversion known as "Reaganomics," or "trickle-down" economics.

Both Thatcher and Reagan, and the many neoliberal politicians to follow them, knew that neoliberalism was about more than just economics. In order to best serve the interests of the economic elite, neoliberalism must be an economic project, but it must also be a political project supported by deep and complete governance. While capitalism is about a mode of production and an accumulation strategy, neoliberalism is also, more importantly, a deep-seated political project as well. It is the political project at the heart of neoliberalism that works to move wealth toward the elite. As with any political project it requires a form of governance to maintain itself. Thus, the focus of this book centers on the transnational flows of capital and labor that converge in the colonias where they are controlled via neoliberal governmentality at the local scale by community leaders and NGOs.

At the heart of liberal and neoliberal economics lies an affinity that ties them to capitalism; all three revere private property and believe it must be protected at all costs. Neoliberalism values private property in much the same way it values privatized resources. Neoliberalism calls for the privatization of many formerly state-owned enterprises. In this model, the state is removed from social service provision, such as the owning of utilities and the supply of public housing, the argument being that these services can be more efficiently run through private, and therefore competitive, companies. In the end, neoliberalism first calls for the shrinking of "big government" and represents a move away from the social safety net of the Keynesian era when big government was the goal (Conway and Heynen 2006; Peck and Tickell 2002).

Without this safety net, a great deal is left up to the individual. There is a great stress on the individual and on individual attainment in neoliberalism. Freedom of choice, much like the freedom of the market, is another central tenet of neoliberal ideology. Values such as self-reliance and self-sufficiency are highly regarded, and there is a general distaste for all collective endeavors as seen in the concept of the greater good put forth in the Keynesian model.

When the state moves toward an individual-centered model and away from the collective and state-sponsored model, much of its service work falls to the third sector (Conway 2006; Wolch 1989). The third sector is made up of private or nonprofit organizations that must pick up where the state leaves off and provide services for the citizens (Jessop 2002). They often work in what is called "partnership" with the state and in this "partnership" pass on neoliberal forms of governance to their clients. This is the case in the colonias and many other low-income communities in the United States and abroad.

Neoliberal Hegemony in the Colonias

Neoliberalism was the guiding political project and economic ideology in both the United States and Mexico in the late twentieth century and the early twenty-first. This means the colonias in New Mexico were developed under increasingly neoliberal conditions. It also means that many of the Mexican immigrant families in the colonias left Mexico because of the neoliberal reforms in that country only to be met in the United States by a new set of hardships generated by similar economic and political ideologies.

The United States and Mexico are more than just geographic neighbors. Their economies are deeply enmeshed and both globalization and neoliberalism have played a big role in this economic integration. Through U.S.-dominated organizations such as the World Bank and the International Monetary Fund (IMF), the United States has exerted an enormous influence over the development of neoliberal economies in the global south in general and in Mexico, in particular. In Mexico, in response to their inability to repay development debts in the early 1980s,[12] structural adjustment programs calling for neoliberal reforms were installed by the IMF and World Bank (Demmers 2001; Martin 2005). Since then, the United States and Mexico have strengthened their mutual commitment to neoliberal economics through free trade agreements such as the North American Free Trade Agreement (NAFTA), based firmly on the tenets of free trade that underscore neoliberalism's aversion to market regulation.

In both countries, neoliberalism makes life harder for the poor as it redistributes wealth toward the rich. Harvey points to redistribution of wealth as a key function of neoliberalism and one of its goals in reversing the equality achieved during the era of Keynesian economics prior to the rise of neoliberalism (2005). Neoliberalism also brings changes to the daily practices of social reproduction in impoverished communities like the colonias. As the state retrenches itself and pulls out of social service provision, someone must "pick up the slack," so to speak. In poor communities across the globe, women and children are picking up the slack as social reproduction begins to encompass more and more practices and services that used to be provided by the state. Isabella Bakker calls this process "gendered restructuring" (1994).

While Bakker points to intensification in domestic labor and individual household responsibilities, my research points to an increase on the level of the community-scale work as well. The women who lead the colonias and their allies are acting on behalf of both their individual households and their communities as a whole when they petition for paved roads or organize for natural gas lines. In their roles as mothers and caretakers, colonia leaders include the welfare of their communities in their daily responsibilities. When the state will not or simply cannot prosecute colonia developers and force them to supply the basic and necessary infrastructure or do so themselves, colonia activists take on the functions formerly performed by the state and supply these necessities for their communities.

The connections between the state's neoliberal policy agenda and the NGOs that serve the colonias are disturbing but not all powerful. Based on an ethnographic inquiry into the concrete situated practices of daily life in the colonias, my research examines the interrelations between communities, the state, NGOs, and social change. I argue that progressive, politically conscious social change that questions systems of power in the colonias is limited, and these limits often appear to be imposed by the relationships between the colonias and the NGOs that serve them. In order to acknowledge, name, and rework these limitations, both colonia residents and NGOs need to see and understand the processes of marginalization and the ways in which the state, NGOs, and communities intersect to create and maintain marginalized communities like the colonias. This is done in part through a careful attention to relations of power and the production of the neoliberal hegemony and the role of NGOs in this process.

Coercion and Consent: Gramsci in the Colonias

In the discussion of neoliberalism and the partnering system to follow in chapter 2, it will be clear that this is not a coercive system. Rather, it is part of the greater neoliberal hegemony. The institutions and processes of the neoliberal hegemony are highly entrenched, as are its discourses, making hegemony central to my analysis.

There are few notions as powerful as the idea of self-sufficiency or "the feeling of pride in a job well done," and it is very difficult to argue with the merits of personal responsibility, especially in the era of "welfare moms" and "illegal immigrants." It is also terribly easy to take part in the daily solidification of the neoliberal hegemony. For example, we take part in this solidification, using a series of powerful discourses that have been used to create consent for and underpin this hegemony, as I just did. For Gramsci, hegemony was about power and the ways in which those with power hold onto it (1989). In a hegemony, the powerful create discourses, or ruling ideas, that naturalize their power and make the inequalities in power and privilege that often exist appear appropriate, even commonsensical. In the current neoliberal hegemony, those with power, who benefit from structures of neoliberalism, live one set of daily situated practices and those who receive inequalities from neoliberalism's processes have a different set of daily situated practices that make these inequalities appear taken for granted as well (Pred 2000). The acceptance of inequality and the taken-for-granted nature of the acceptance are crucial to the consent with which hegemony is created. Raymond Williams describes Gramsci's theory of hegemony and the devastating ways in which it is inscribed on those without power. For Williams, hegemony is "a lived system of meanings and values . . . It thus constitutes a sense of reality . . . If the pressures and limits of a given form of domination are to this extent experienced and in practice *internalized*, the whole question of class rule, and of opposition to it, is transformed" (1977, 110, emphasis in the original).

As a political project, neoliberalism is exceptionally skillful at creating hegemonic power. Katharyne Mitchell approaches hegemony from a spatial point of view in her work on neoliberalism and Chinese migrants in Vancouver (2004). She brings space into the equation when analyzing hegemonic powers and points to its foundational place alongside history in any discussion of hegemonic formations. Mitchell discusses the common privileging of the historical over the spatial in studies of hegemony and her desire to undo this focus. While she acknowledges the role of history, "hegemonic formations are also made in the production of, movement through, and representation of space . . . Hegemony is made not just in time, but also in space—in the taken-for-grantedness of bodies in specific places, what those bodies are, what those places are, and how they constitute and define each other" (17). The lived spaces of hegemonic neoliberal power is a theme central in the study of the colonias as they are essentially one such embodiment of these spaces. Throughout this study, I will pay attention to the spatial as well as the gendered aspects of daily life in the colonias, especially as they pertain to the development of leadership and community activism.

The key to hegemonies is that they are never fully complete. There is always room for resistance, or counterhegemonies, to develop; hegemonic powers must be ever vigilant and maintain some sort of system to react to the possibility of revolution. Here is where the neoliberal hegemony excels. Like other hegemonies, neoliberalism must find ways to encompass or co-opt these counter and alternative hegemonies and make them part of itself to neutralize them as threats. Lisa Duggan expertly analyzes this process as the neoliberal machine moved in on the gay rights movement, as seen in the online writers' group, the Independent Gay Forum (IGF): "The new neoliberal sexual politics of the IGF might be termed the *new homonormativity*—it is a politics that does not contest dominant heteronormative assumptions and institutions, but upholds and sustains them, while promising the possibility of a demobilized gay constituency and a privatized, depoliticized gay culture anchored through a double-voiced address to an imagined gay public, on the one hand, and to the national mainstream constructed by neoliberalism on the other" (Duggan 2003, 50). In Duggan's example we see neoliberalism's amazing ability to move into social movements, particularly those based on identity politics, and take a solid hold in order to supplant progressive groups with a "demobilized" or "depoliticized" variant. This is a political project, which uses its powerful governmental techniques to quell or encompass its opposition, at work.

Neoliberal Hegemony in Mexico

In Mexico, the tenacity of neoliberalism has played out through several political parties, but one constant has remained: the state's focus on emigration to the United States and the income it earns from remittances from the United States. Neoliberalism in Mexico is based on a similar ideology to that in the United States (Martin 2005). Economic austerity through the regulation of state spending, a strong stress on free trade and the deregulation of markets, and a push toward privatization reshaped Mexico from the 1970s onward. One important aspect of the neoliberal project in Mexico has been the repeated devaluation of the peso. After each round of peso devaluation, employment in Mexico fell and the country suffered severe economic crisis. In these situations, migration across the border increases. The most notable of these devaluations took place in the mid-1970s, after which the economic crisis was so severe Mexico went to the World Bank and the IMF for help and began a long and often difficult relationship with these global lenders. The next was in 1982, when Mexico defaulted on their loans to the World Bank and IMF, at which time they took on a neoliberal system in an effort to repay these loans. Yet, in 1994, they were forced to devalue again, this time in light of the events in Chiapas with the Zapatistas and the beginning of the NAFTA era. In all of these episodes of devaluation, the effect on the border region was large, particularly so on investor confidence in the maquiladora sector. This led, in both 1982 and 1994, to an out flux of companies to other international low-wage production areas (Esparza and Donelson 2008). This meant many people were looking for new and better paying jobs and made the move across the border into the United States and into colonias.

Although the economic and political tenets of neoliberalism were applied differently in Mexico, they were done so in an equally hegemonic style. According to Adam Morton, neoliberalism in Mexico can be seen as an accumulation strategy that protects the capitalist class yet was "pursued while reconfiguring the hegemonic project of the PRI" (2003, 636). This resulted in Mexican neoliberalism being defined as an accumulation strategy, similar to what Harvey discusses in the United States, and a hegemonic project at the same time (2005, 19). In Mexico, as in the United States, it is the economic elite who accumulate the most wealth in the neoliberal system.

The disparity in wealth means those who do not benefit must often look elsewhere for sources of income for survival. For many, this involves migration, both seasonal and more permanent, to the United States. The Mexican state encourages its citizens to make the journey to the United States, as it knows the remittances they send home are crucial to their economy. The transnational nature of the colonias means there are both cultural and economic ties between these communities and Mexico. These economic ties provide by some estimations 2.8 percent of the Mexican gross domestic product, which was approximately $13 billion in 2006.[13] This is below the level of oil exports yet above revenues from tourism and agriculture. With numbers like these, one easily sees why the Mexican government is eager to negotiate a new guest-worker program with the United States. In the meantime, the colonias serve as a way station for immigrant labor that desires to own land and not "waste" money on rent while they live in the United States. Jobs in the United States have become one of the most popular solutions to poverty in Mexico, and this does not seem to be shifting, neither within the colonias nor with the help of the Mexican government, any time soon. Because the United States wants cheap labor and Mexico needs jobs, the colonias will continue to have a steadily growing population of residents.

Neoliberalism, Capitalism, and Globalization

The tenets of neoliberalism can be seen at work within the colonias alongside capitalism and the processes of globalization. The three are intricately tied together in many spaces and cannot be easily, or usefully, extricated much of the time. Neoliberalism as a political project would fail to exist as we know it without capitalism. Just as capitalism leads to uneven development, the path of neoliberalism has proven to be uneven as well (Harvey 2005). In this section, I want to examine why, when focusing on the gendering of activism in the colonias, the term "neoliberal" best describes the hegemonic nature of the social, political, and economic processes at work. I choose to stress neoliberalism because of its social history in both the United States and Mexico. In the recent past, neoliberalism has shaped the lives of the poor and women in very specific ways in both countries. For the purposes of this research, I stress the particular forms of governmentality that accompany neoliberal regimes. Finally, I wish to discuss neoliberalism's complicity in programs of "development" in the global south, its relationship to NGOs and women leaders there, and the parallels I see to NGOs and leaders in the colonias (Li 2007; Rankin 2004). For all of these reasons and others, neoliberalism is the primary process in the colonias that shapes the gendering of activism and its limits.

The rise of discourses of individual responsibility and the growing stress on individual self-sufficiency and self-reliance that accompanies neoliberalism's popularity in the nation's imagined community help create the neoliberal hegemony. This hegemony creates the consent necessary to fuel the cutbacks to social services and the state's retrenchment that affect women and children across the social spectrum. This is where we see women and children "picking up the slack," or partnering with NGOs to make ends meet, as the state moves away from service provision. According to Jean and John Comaroff, neoliberalism simply intensifies abstractions already inherent in capitalism: the separation of labor power from its human context and the replacement of society with markets (2000, 305). They argue that these processes may not have been completely successful, but they have come a long way, usually unevenly, across space. From this process, Comaroff and Comaroff continue, "emerges a more radically individuated sense of personhood . . . In this vocabulary, it is not just that the personal is political. The personal is the only politics there is, the only politics with tangible referent or emotional valence" (305).

As politics is individualized, so too is each individual's relationship with the state. When the relationship that used to exist between the collective "social good" or "social body" and the state in the Keynesian era ceases to exist, it is replaced by a neoliberal version. In this version, the relationship between the state and the individual is the most important relationship. Social problems, and their solutions, can no longer be seen as structural but rather must be seen as individual, and so too must their solutions. Thus, in the colonias, we will see that activists are taking charge and working as individuals with neoliberal solutions to lead their singular colonias out of resource deprivation. As Margaret Thatcher famously said, there is "no such thing as society, only individual men and women" (Harvey 2005, 23).

It can be said that the cultural politics of neoliberalism are, in fact, those of the "market culture" and that the overall project of neoliberalism may be to transform the many diverse cultures of the world into market cultures (Rankin 2004). This project has met with varied success, but where it has been most successful you will find increased personal responsibility and individual self-reliance to be the hallmarks of the neoliberal citizen. The real power of neoliberalism lies in great part in its ability to create consent, which translates, in Gramscian terms, into its ability to create and maintain hegemonic power. In democracies such as the United States or the United Kingdom that value "freedom" and "independence" fiercely, the neoliberal focus on the individual and personal responsibility is an easier sell, so to speak. Creating consent for a system that holds the individual responsible for her own well-being, removing the state from the equation, and changing the whole idea of collective well-being is not as difficult as it might be in a culture that is not as centered on the individual in the first place. It becomes easier to see how neoliberalism became "natural," to see how Thatcher was able to sell her "there is no alternative" campaign, or, as Wendy Larner puts it, to "explain why some people (sometimes) act as neoliberal subjects" (2003, 511). Larner is expressing the fact that, in many instances, people are not forced into their neoliberal lives but rather willingly consent to take on these roles and simply choose to live in this market-driven reality as it has been naturalized.

Powerfully connected to theories of the political economy of neoliberalism are theories that address how neoliberalism, as a political project, uses specific forms of governmentality to pursue its goal of a minimalist state in the service of capitalism. Michel Foucault, in his studies of power, developed a theoretical model of government as the "conduct of conduct." In this understanding of governing, the primary processes at work are those that shape or guide the conduct of a person or a people. There are two central points to be made here. First, when Foucault made the move from government guiding the conduct of a person to a *people*, it signified the shift to governing a population. This is a very important shift in Foucault's theory, as it also signifies the move to the governing of an economy

(Foucault 1994, 217). In Foucault's theory of government, as articulated in his lecture, "Governmentality," the goal of government is "not the act of government itself, but the welfare of the population" (216). As a governing system grows, its knowledge is tied to knowledge of its population. Foucault describes this new knowledge as political economy: "the new science called 'political economy' arises out of the perception of new networks of continuous and multiple relations between population, territory, and wealth" (216). Second, as we focus on governing as the "conduct of conduct," it is important to think of governing as a multiscale process, moving from the governing of the self to governing practices within social institutions and those between the state and individuals. Governmentality in the colonias reaches from the scale of individual leaders, through their interactions with local institutions, to the state. It is constantly tied to their place in the regional economy. For these reasons and more, Foucault's theories of governmentality and their many interpretations fit well with the ethnography presented here.

In Mitchell Dean's articulation of the ties between liberal government and authoritarianism, it is clear that the neoliberal state, following from the liberal state, governs through a "set of binding obligations" rather than the freedom that is commonly understood as central to the democratic state (2002). Moving from the Foucauldian idea "that the essential issue in the establishment of the art of government" is the "introduction of economy into political practice" (Foucault 1991, 92), Dean uses Foucault's theories of the state's role in the regulation of daily practices to tie political economy to governmentality. In particular, Dean highlights the role of civil society in these processes: "The term 'civil society' can serve as convenient shorthand for all that liberal government must take into account that is exterior to the formal governmental domain of the state" (2002, 40). It is important to note the phrasing Dean uses because "exterior to" is not the same as "separate from," and it is the inescapable connections between civil society and the state that play a large role in the disabling of activism in the colonias.

In the colonias, the ties between civil society and the state closely follow the pattern Dean spells out: "A liberal approach to government generates specific norms of individual and collective life and hence of forms of freedom that become the means and the objectives of liberal governmental programs. In such fashion, the liberal government of civil society can specify the kinds of freedom and autonomy that are to become the objectives of governmental policies and practices which, under certain circumstances, will require the use of instruments of coercive authority and legal sanctions" (40). In the colonias, these "norms of individual and collective life" are shaped by both NGOs and the state to reflect neoliberal ideologies of independence and self-reliance. Foucault similarly stresses the importance of "disposing things" to the state's benefit in the art of governance: "With government it is a question of not imposing law on men, but of disposing things: that is to say, of employing tactics rather than laws, and even of using laws themselves as tactics—to arrange things in such a way that, through a certain number of means, such and such ends may be achieved" (1991, 95). My research demonstrates how colonia communities and leaders are disposed by the state to deal with their resource deprivation on their own, in classic neoliberal fashion, thereby often limiting progressive social change that challenges systems of power.

Dean also develops a theory of "liberal police" who work to "guarantee individual liberty" through civil society. He takes this theory a step further when he writes, "Today what I shall call 'liberal police' is exercised throughout much of the world in some relation to a pervasive governmental rationality and perception of economic necessity, that of globalization" (Dean 2002, 41). Thus, neoliberalism can be understood to be related to global economic restructuring, the growing territoriality of capitalism, and the production of self-sufficient citizens. Colonias, as working-poor communities, are particularly hard hit by the retrenching of state-provided social services and thus turn to NGOs, which pickup these services without understanding how governmental these "non-governmental" organizations can really be (Morris-Suzuki 2000). As the state pulls out of social service provision, it is usually the women who are hardest hit and who must make due by performing more tasks themselves while looking to outside organizations, such as NGOs, for help. When these organizations employ individually focused self-help techniques, they reinforce already powerful neoliberal discourses and often lose sight of the larger-scale processes that could be used as a rallying point to unite marginalized communities. This project acknowledges that local engagements with neoliberal ideology are not automatic nor are the consequences of neoliberal policy the same in every place. My research presents a much richer picture that neither theories of political economy nor governmentality alone can explain. Only a materially grounded account of daily specialized practices can really get at the multiple and interconnected relations between grassroots activism and the growing influence of neoliberalism.

Neoliberalism Here and There

The consent neoliberal social policy creates not only is present in the United States but can also be seen as a result of the actions of U.S. institutions across the globe. Particularly in the global south, the role of neoliberal economics and social policy has had devastating effects on women and children, as many of the organizations and institutions that formerly provided public services and aid have been privatized or simply disbanded. In these circumstances, NGOs develop to meet the needs of communities that cannot access the often more-expensive privatized services or of communities for whom privatized services simply do not exist. Under these conditions, several themes arise that are also apparent in the colonias that are similar to those in developing countries with extremely limited resources, overworked populations, and a great reliance on NGOs for basic services and interactions with the state.

The first of these themes is the uneven nature of neoliberal development. According to Larner, particular spaces and institutions produce neoliberal policies and in turn these policies produce spaces. Larner explains, "It may be more useful to think of particular spaces, states, and subjects as artifacts, rather than as architects, of neoliberalism" (2003, 511). This helpful interpretation applied to daily life in the colonias views neoliberalism as a force that produces social spaces where women leaders respond with their activism to the resource deprivation created by neoliberalism.

Although neoliberalism may have very clear intellectual roots, it manifests itself in different ways in different places by interacting with local politics and cultural practices, creating regionally unique outcomes. According to Jamie Peck and Adam Tickell, the "extralocal" forces of neoliberalism may be "coercive" and strong, but nonetheless, their "outcome is not homogeneity, but a constantly shifting landscape of experimentation, restructuring, (anti)social learning, technocratic policy transfer, and partial emulation" (2002, 396). Neoliberal theory clearly dictates the economic, political, and social changes that are to be implemented under a neoliberal regime, yet we see a vast array of different types of neoliberal systems at work in the world today. This variety is a testament to the uneven development of neoliberal policy and the variability in neoliberal governments. It is also a testament to the power of neoliberalism to adapt to its environment (Duggan 2003).

In both the global south and north, the growing influence of the neoliberal political project has been accompanied by an increasing interest in, and fetishization of, civil society (Comaroff and Comaroff 2000, 330). "More aspiration than achievement": these are the words Jean and John Comaroff use to describe civil society as it grew in popularity around the millennium. Civil society became the wonder cure for all that ailed neoliberal society. What is most fascinating about this mounting interest in civil society is that it encompasses both those who support the processes of neoliberalism and those who fight to staunch its tide. Advocates of neoliberalism see the elusive realm of civil society as the perfect replacement for the services of the state that neoliberalism deregulates and shuts down. Critics of neoliberal movements see civil society as a place to regroup and a platform from which to mount their attack, a place to create a counter-force, a space in which to build a "people's army," or a counterhegemonic democratic movement. In this way, civil society becomes a contradiction-laden space in the neoliberal era, especially so in the colonias. In the colonias, the primary representatives of civil society are the many NGOs that offer services to colonia residents.

Transnational Border Communities

Nearly everyone in the colonias of Los Montes and Recuerdos is Mexican. For all intents and purposes, they are Mexican communities. Yet these are in the United States, and many of these Mexican immigrants are U.S. citizens, and others are fully documented residents with social security numbers and all the other symbols of "American-ness." At the same time, you often feel as if you have stepped across the border when you walk into a colonia: the houses are often surrounded by brightly painted flower beds, and there is loud Mexican music playing on a sunny Saturday morning. It always smells like a strong chile is cooking, and there are other trappings of Mexican culture on display. Colonias come to exist out of transnational flows of labor and capital that converge in isolated pieces of unused farmland. They are communities with significant ties to Mexico that are reproduced in the United States. In their physical isolation, it is that much easier to produce these dominantly Mexican spaces.

Transnationalism is a set of processes often generated by the global restructuring of capital. On the individual scale, these processes can be examined in the ways immigrants' lives develop to sustain social relations that transcend borders through familial, economic, political, religious, social, community, and other connections. These connections and processes often demonstrate how individuals, families, and even whole communities are involved simultaneously in two or more countries (Glick-Schiller, Basch, and Blanc-Szanton 1992). If transnationalism is the set of processes by which immigrants manage their lives in two or more nation-states, then it becomes necessary to acknowledge that these processes have both visible and invisible geographies. While the visible geographies of labor supply and demand attracted immigrant laborers across the border and created colonia communities in the first place, the daily geographies of these Mexican immigrant communities are tied up in cultural and political geographies that can be less visible but equally transnational in nature. It is often the invisible geographies, such as the negotiation of gender roles and relations in the household, where some of the most interesting and overlooked dynamics in transnational communities take place. This is the case in the colonias where only a few women negotiate their place in a new transnational community and familial system to become leaders.

Aihwa Ong creates a nuanced and easily applied definition of transnationalism useful for describing the colonias. Writing about the transnational practices of Chinese subjects, she is "concerned with transnationality—or the condition of cultural interconnectedness and mobility over space— which has been intensified under late capitalism" (Ong 1996, 4). Colonias, too, are culturally interconnected. Many residents move into these Mexican communities *because* they are all-Mexican and sustain strong ties to Mexico. However, at this point the comparison becomes a bit uneasy. Mobility is not something late capitalism brought to the colonias. Colonia residents, on their average $5,000-a-year incomes,[14] have limited mobility and experience a limited form of "time-space compression" as an effect of global economic restructuring (Harvey 1990). More often, they experience an inability to afford gasoline and face harassment at the hands of the border patrol. It does not seem to matter that the majority of colonia residents are documented residents or citizens. Yet, within these all-Mexican spaces, colonia residents can move with ease, and this is part of their attraction to Mexican immigrants. Immigration from Mexico to the United States and remittances from immigrant communities in the United States has been a sustaining aspect of the Mexican economy for decades and in many ways is a defining aspect of many transnational Mexican communities on the border.

Transnationalism, alongside neoliberalism and globalization, shapes leadership in the colonias. As transnational communities, colonias have particular needs and desires aside from basic infrastructure that shape the work of colonia leaders. Although colonias are clearly transnational communities in many ways, I believe it necessary to qualify this statement. After working in these communities for eight years, my initial response when asked about community organizing in the colonias would not be to respond with a discussion on transnational issues but rather to comment from a local-actor perspective. Like Ana Aparicio in her work on Dominican Americans, I find colonias, though inherently transnational, must first be addressed on the level of local relationships and activities (2006).

The needs of border communities, especially in the late twentieth and the early twenty-first centuries, have been attuned to the blatant anti-immigrant rhetoric of the United States, and colonia leaders have to be aware of this as well. Women leaders also have another set of issues with which they must deal. This involves the often-conflicting gender roles they encounter when they arrive in the United States. These roles can set up expectations that are in opposition to those with which they have lived in Mexico, and reconciling transnational differences can put a great deal of stress on relationships.

Development of Transnational Gender Systems

One of the most complex and important sets of social relationships that colonia residents negotiate between their origins in Mexico and their current lives in the United States are gender relations. For women leaders in particular, the transnational and domestic gender systems in which they live both enable and disable their leadership. When these transnational gender systems are exposed to neoliberal governance, they result in women leaders who often act as neoliberal political subjects.

That leadership in the colonias is gendered is not particularly surprising. In other communities around the world, women are the central activists and leaders (Gilmore 2007; Kaplan 1997; Naples 1998a; Pardo 1998a, 1998b). But the ways in which gender shapes leadership and social change in the colonias is very particular. Colonia leaders experience globalization and neoliberal policy as Mexican women, as mothers, and as the poor. Their activism is shaped by the ways their gender and ethnicity intersect with larger-scale processes. One such process is the development of transnational gender systems in which colonia women must navigate "traditional" Mexican structures of patriarchal power, or machismo, in a new country.

"Machismo," as it is constructed in both Mexico and the United States, is a fluid and ever-changing structure that develops and redevelops at different scales as it intersects with other gender systems. According to Gutmann, machismo grew in popularity as a concept during the Mexican revolution when it was associated with courageous actions on the part of both men and women, "though the terms used to refer to courage carried a heavy male accent" (1996, 124). It was not until the 1940s that "Mexico came to mean machismo and machismo to mean Mexico" (124). From this point on, machismo became a common framework for gender relations in Mexico. Although still a strong influence in the Mexican families with which Gutmann works, "Mexican working class men as well as women, have learned to manipulate the cultural rituals and social laws of machismo" (3), as have the women and men in the colonias. It is in the family that colonia residents "skirmish over gender identities and relations" (256). These skirmishes can come to have lasting effects on a woman's ability to take part in the community life of her colonia. Throughout this work, I focus on variations in gender relations in the home as an important site for struggles over women's community activism.

Ethnography

By studying a particular historical and geographical conjuncture, this project examines how colonias developed in New Mexico as Mexican immigrant communities led by women and NGOs, which work with women activists to shape, yet ironically limit, social change in the colonias. When globalization, neoliberalism, and transnationalism are the focus of study, we must be careful to examine the framing of discourses surrounding these concepts. As we have seen, it is well within the agenda of the neoliberal political project to produce discourses of the inevitability of neoliberal policy and of its dominance in global markets (Hart 2002). These discourses can be described as "impact models," which describe the impact of the "global" on the "local" as inexorable and all powerful. In an effort to reject impact models of globalization and social change, I use an approach that focuses on the points where large-scale processes come into contact with particular relations and institutions. This research uses situated daily practices to demonstrate that the local and the global are not the only meaningful scales and that connections between all scales are important. This study also illustrates the ways in which the local scale always has a measure of autonomy and, in turn, shapes the scales with which it interacts, leading to the ethnography of convergence. Using ethnographic detail that speaks to the material grounding of the multiple-scaled processes involved in the colonias and the life stories of eight women leaders, this study argues that the relationship between women activists and NGOs is such that progressive social change is limited in the colonias. My arguments surrounding the limitations to activism in the colonias reject the unidirectional impact-model assertions that are the foundation of the neoliberal project and are based rather on the complex interrelations between regional and global processes, which I observed daily in three colonias.

This research seeks out, acknowledges, and focuses on the multiplicity of contradictory and ambiguous relationships that exist in and around colonias. Contested, contradictory, and conflicting situations may be hard to interpret, but in fleshing out these complexities, a great deal can be learned about both the powerful and the marginalized (Moore 1998). It is only through exploring complex and contradictory situations that the simultaneously enabling and disabling nature of women's community activism and leadership in the colonias can be understood. Most importantly, these interconnected and entangled relationships can point to cracks in the power and hegemony of dominant groups. It is through these cracks that alternative and subordinate relationships may develop because relations and processes of power are "restlessly being reworked" (Nagar et al. 2002, 267). This is especially true when the case in hand, like this one, is a set of relationships based on people's daily practices and not the faceless corporations and financial flows that so often characterize studies of "globalization."

If this is a story of the production of a place as a site of convergence, then this is an ethnography of convergence. An ethnography of convergence is a story of a site of conflux, in this case the growth of a community and its leaders and the processes that take place as this community grows and evolves. The processes documented here are those of neoliberal governance in the era of flexible globalization on the U.S.–Mexico border. Like a critical ethnography, an ethnography of convergence focuses on "processes of constitution, connection and dis-connection, along with slippages, openings, and contradictions, and possibilities for alliances within and across different spatial scales" (Hart 2006, 982). The important similarities here are the possibilities for contradiction, openings, slippages, and connections. There are always multiple possible ways in which a set of processes can come together in the social production of a place. Therefore, there are multiple possible ethnographies of convergence.

Since this research examines "global" processes and institutions as they intersect with daily life, ethnography was the obvious method for this study. By grounding this study in the details of daily life while acknowledging that these practices are embedded in relations that are tied to the global scale as well, this book takes on a complex task.

The relationship between the local and the global and between micro- and macrosocial relations has been the subject of much discussion in the social sciences. Ethnography, with its emphasis on the detailed fieldwork of daily, situated practices can help further dissolve the false dichotomies between the local and the global, micro and macro. For example, Cindi Katz discusses the advantages of ethnography in her work on globalization and the daily lives of Sudanese children: "It is more interesting to me to examine the intersecting effects and material consequences of so-called globalization in a particular place, not to valorize either experience or the local, but, quite the opposite, to reveal a local that is constitutively global but whose engagements with various global imperatives are the material forms and practices of *situated knowledge*" (2001a, 1214). Katz's stress on the local as "constitutively global" is closely related to Doreen Massey's argument about place that seeks to "problematize the distinction between the local and the global; if each is part of the construction of the other then it becomes more difficult to maintain such simple contrasts" (1994, 9). The "simple contrasts" Massey speaks of are those that oppose the local and the global in such a way as to obscure the intricate relations between the two. The ethnographic data that support my research not only focus on the details of daily practices in these communities but also allow the interconnections that exist between these localities and the global processes of economic restructuring and neoliberalism in which they are directly involved to be revealed. As Massey argues, "At minimum we can say that localities are not internally introspective bounded units. They have to be constructed through sets of social relations which bind them inextricably to wider arenas, and other places" (142). In the colonias, these social relations stretch out from the colonia to the Catholic Church, to local NGOs, to county offices, and across the border to their family histories in Mexico. The very existence of colonia communities is a testament to the importance of relationships between localities and global-scale processes. If it was not for economic integration between the two countries, immigration from Mexico to the United States would not have led to the need for colonia communities. In order to understand the development of the colonias, you must examine these multiscale interactions.

As an ethnographer you live through many of the same processes and structures that your informants do, although your experience can never be exactly the same (Haraway 1988; Visweswaran 1994). Yet, as Michael Burawoy et al. point out, "global ethnographers cannot be outside the global processes they study" (2000, 4). If globalization is defined in part by "the recomposition of time and space—displacement, compression, distanciation, and even dissolution," then the job of the ethnographer is to "shed light on the fateful processes of our age" by sharing in these processes as they are lived everyday (4). My worries as a white, middle-class woman with a postgraduate degree were very different than those of the colonia residents with whom I was working. After the 2000 presidential election, which occurred while I was in the colonias, I found my thoughts and fears moving to scenarios of what will happen to "women's rights" as a catchall for wage equality, reproductive freedom, and freedom from violence. These concerns were also on the minds of the women leaders but so, too, were issues of immigration and daily racial and ethnic discrimination of which I had no firsthand knowledge. In the end, we have all lived through some of our worst dreams and have discussed these nightmares and the challenges they present to organizing in the colonias in this extended current moment of danger (Pred 1995, 1084).

Methods

This ethnography was conducted in three colonias in Doña Ana County, New Mexico between January 2000 and December 2000 with follow-up trips in 2001, 2002, 2005, and 2007. The research began in earnest in 1998 when I was first introduced to the Community-Organizing Group (COG). I was in New Mexico doing early research on the colonias and was told by several local scholars that the COG was the most important NGO working in the colonias. The COG director, Elena, was unavailable during my first visit, so I met with Ernie, one of the community organizers. Ernie and I got along well and he was interested in my work. Several months later I began the negotiations to work with the COG in their member colonias. I use the word "negotiations" here because before I could begin my work I had to prove myself both theoretically and practically to the COG. They wanted to know about the theoretical foundations of my research and how I proposed to both protect the colonias and give something back to them. The colonias and the COG had a history of bad experiences with researchers who came to study and then disappeared, leaving behind only empty promises and bad feelings. Once I agreed to never make any promises I would not keep and to always appraise the COG of my work, I was able to begin the study.

My daily practices while in the field were dictated by the schedules of the women in my sample and the COG organizers with whom I worked. The first three months of my fieldwork were spent shadowing two COG organizers, Mario and Ernie. I attended COG staff meetings and community visits with the organizers, and it was in this way that I met the eight women who would be the core of my study sample and was introduced to many other NGOs that worked in the colonias. I also spent a fair amount of time working with Sarah, the economic development coordinator COG had hired just before I came to Doña Ana County. Sarah's job was to develop job opportunities in the colonias using projects that would advance and benefit the communities. She and I shared an interest in women's issues and both believed a feminist approach to community development would serve the COG if we could move the organization in that direction. As the year progressed, I would be mistaken for Sarah on more than one occasion and, although I always corrected these moments of mistaken identity, I also took advantage of them to dialogue with colonia residents who might not have otherwise engaged me in conversation. I did not feel badly about this, as I was usually able to answer their questions as well.

When I was not with the COG organizers, I was interviewing local county officials and others involved in issues relating to the colonias. Inevitably, my connections to the COG opened many doors for me while shutting others. Almost all the county and state officials involved in the colonias respected the COG and, therefore, welcomed me into their offices. The developers, on the other hand, had no time for the COG, or me for that matter. Therefore, my firsthand experience with developers is limited. Most of my information on colonia developers is from the colonia residents who bought land from them and from the county and state lawyers who prosecuted them.

During the first few months, I spent time piecing together the histories of each colonia through lengthy discussions with past and present COG organizers and county staff. During this period, I also established relationships with the women leaders. Ironically, at the time, I was not aware that this was happening—I thought my work was failing and that I might never really get to know the women leaders. I soon realized that the most important thing I had done to prove myself to the women was to stick around. The women, unlike the COG, had no interest in my theoretical grounding. They simply wanted to see if I was serious; would I stay if they ignored me? When I kept coming back every day for several months, they decided I was serious. Then they started to return my phone calls and even called me when they thought there was something going on that would interest me.

The turning point came one day when Estella, the primary leader in the colonia of Recuerdos, confronted me about my dedication to my research. I had been in Doña Ana County for about four months at this point, and I was experiencing frustration regularly as leaders would fail to alert me to events in the communities no matter how often I reminded them that I was there to study community organizing. On this day, Estella took me aside and, loud enough for the others present to hear, said with a very serious tone, "You're a real *cabrona*, aren't you?"[15] This could have been interpreted as quite an insult, but I realized fairly quickly that she was actually initiating me in a way into the community and that the right reaction was absolutely necessary. I had to be tough and shake this off. She had called me a bitch, but in a good way. So I agreed, "Yes, I am." Then she smiled and laughed and went on to explain that she realized I was going to stick around. After all, I had made it to that meeting all on my own without anyone telling me about it, so I must really be interested in their community. From that point on, I usually got a phone call when something happened in Recuerdos. News has a way of moving from one colonia to another in the communities in which the COG works so my initiation with Estella was soon common knowledge among leaders in the COG colonias, and this helped pave my way into the community organizing networks in the other colonias as well.

The bulk of the descriptive material used in this work comes from observations at COG staff meetings, community meetings, and my daily interactions with leaders in their homes and at work. The majority of the quotes come from a series of four in-depth interviews I conducted with each of the eight women in the core sample. The topics ranged from domestic gender systems and the roles of men and women in the home to their feelings on the contested 2000 presidential elections that occurred while I was in the field. I conducted these interviews in Spanish and all of my other interactions with the women leaders were in Spanish. Only one leader, Alicia, spoke English, and she and I spoke English when we were alone and Spanish when other leaders were present, as colonias are Spanish-speaking communities. The rest of the quotes come from my field notes. I always carried my small blue notebooks and spent many hours sitting in my truck on the side of the road, just outside a colonia, frantically scribbling notes on the day's events.

Questions of Method

The empirical focus of this project is a group of eight women who are leaders in their colonias and the multiple relationships that exist between them and their communities. As the following chapters will make clear, the women's self-identification as leaders varied from those who actively label themselves "leaders" to those who simply chose to consistently attend "leadership" trainings and meetings. All the women in this study understood that I was investigating women's leadership and agreed to be interviewed as leaders in their colonias.

I choose to use the word activism to describe the work the women are doing in their communities as a political move. Activism itself has been under attack in this country over the last decades and in many circles is seen as a dangerous or subversive thing. Yet these women work to create real and necessary changes in the way they live day to day, to improve their quality of life, change opinions about their communities, and to attain some basic civil rights they are lacking. This is activism. They are trying to create progressive social change. The question that haunts this book is why is the progressive social change that confronts systems of power they create often limited to their individual lives?

Within the debates surrounding locality studies such as this, a series of important interventions have led to new understandings of how case studies can be used to document translocal processes.[16] This ethnography is a locality study at heart, grounded in the daily practices and meanings of life in the colonias; yet it is also deeply engaged with a series of important transnational processes. For example, what it means to be an immigrant farm-worker community produced through neoliberal political policy and exposure to neoliberal discourses can be seen in the daily practices of colonia leaders that exhibit high levels of self-reliance and independence.

Of the over two hundred and fifty women living in the three colonias involved in this project, I found only eight actively identified as leaders. I offer explanations for this very small percentage of women who become leaders throughout this book, so I will only briefly touch on them here. The production of the transnational gender systems that exist in the colonias dispose only a few women to leadership. More specifically, the individual gender systems and gender relations in each household exert a strong influence on which women become leaders. Leaders tend to come from more egalitarian households or single-mother households. Mothering, the highly gendered responsibilities of caretaking, and the discourse surrounding caretaking also dispose women to leadership. All the women in my sample are mothers and took part in similar daily practices as caretakers.

Individual motivations stemming from personal life history and exposure to particular discourses are another important explanation for why some women become leaders and others do not. Personal motivations for leadership are buried in each woman's biographical trajectory. Although all colonia residents are produced to some extent as citizens serving the neoliberal political system, not all women become leaders. For only a small percentage of women colonia residents do their personal life histories combine with the discourses of neoliberal self-reliance and the discourses of caretaking used by NGOs to recruit women such that they become leaders. The COG and other NGOs must be aware of the multiple processes that produce women as leaders in order to capitalize from this production. NGOs learn how to read the circumstances of women's lives and are able to recruit those who have leadership potential.

The names of both the colonias and their leaders have been changed to assure anonymity. Because of the small sample size and the very personal nature of the topics being discussed, I have changed details to protect the identities of informants.

The Colonias

Recuerdos

An hour northwest of Las Cruces, about half way to HUD's 150-mile limit, in an idyllic little valley, sits the colonia of Recuerdos. Recuerdos was declared a colonia in 1993. Recuerdos is a relatively small colonia with approximately thirty-five households, surrounded by hills on three sides and a highway on the fourth. Like most colonias, Recuerdos is made up of a mix of old and run down mobile home trailers and self-built housing (see figure 1.1). Located in the greenbelt that surrounds the Rio Grande as it descends through the state, Recuerdos is set firmly in an agricultural region. Nearly everyone in this area makes their living in agriculture or a related field. Most of the residents in Recuerdos work in the fields picking onions and chile. Chile is the backbone of their economy. Those who do not pick chile in the fields work in the processing plants. Those who do not work outside the home most likely work making ristras during chile season.

Recuerdos lies about a half-mile outside the village of Hatch, an old town with a long history of farming and a strong Anglo community leadership. Recuerdos, on the other hand, is an all-Mexican community. Once you are there, you feel as if you are miles away from anything and anyone. The lots are rather spread out and there is much more open land and green areas than in many colonias. This is in part due to the presence of a great deal of Bureau of Land Management (BLM) land within the boundaries of the subdivision itself. Because of the lack of regulation when colonias are subdivided, they often incorporate land the developers never had legal access to in the first place. In the case of Recuerdos, this BLM land is rented to ranchers for grazing of cattle that wander around Recuerdos on a fairly regular basis. It also means that Recuerdos has a built-in nature reserve. Unlike Valle de Vacas and Los Montes, Recuerdos is not all sand and dirt; it has a more diverse palette than the monochromatic browns that make up the other two colonias. In Recuerdos there are a few trees and a good deal of green brush in the spring.

A May 12, 1994, article in the *Las Cruces Sun News* reported, "A federal judge told the village of Hatch Wednesday it could not enforce a zoning law that Mexican farm workers say was created to kick them out of town . . . The farm workers say the 1990 Hatch zoning ordinance, which was revised in 1993, bans mobile homes and mobile home parks from

Figure I.1 Typical colonia housing stock. This trailer in Los Montes is similar to what would also be found in Recuerdos or Valle de Vacas.

the village and was created because most mobile homes in the village are occupied by Mexican farm workers" (Goldsmith 1994). Although the law in question began in 1973 as a ban on mobile home parks in low-density residential neighborhoods, it was not enforced until 1993 when the Anglo residents of Elm Street complained about all the Mexicans living in the street's two mobile home parks. From the 1994 *Sun News* article and my conversations with people involved in the case, I gathered it took so long for the law to be used against these two mobile home parks because it was only over time that Anglo residents became alarmed about the rising numbers of Mexican immigrants in the community. Furthermore, it was only Anglo residents who complained to the village and requested that the law be enforced. Once the village threatened to kick the Mexican farm workers out of the parks, the farm workers went to Southern New Mexico Legal Services and started a discrimination case against the village. When I got to Doña Ana County, the case had been tried and won by the farm workers, and the village of Hatch was on probation with a panel made up of local church and legal professionals watching its every move. Several of the people on the village's "probation committee" were also involved with the local colonias. They told me that during the 1993 enforcement of the law, these colonias, and Recuerdos in particular, were considered to be the village's solution to the "Mexican problem." The idea was for the Mexican farm workers to move out of the village, where they were not wanted, and into the colonias, where they "belonged."

Los Montes

Like many colonias, Los Montes is located off the main road behind agricultural fields. Before the county negotiated a proper access route, colonia residents had to either illegally cross the field between the highway and the colonia or cross a very old and rickety bridge over an irrigation ditch. Neither option was acceptable. Now, thanks to a lengthy and successful lawsuit brought against the colonia developer, residents can drive up a newly paved road through the field they used to have to cross illegally.

Once inside Los Montes, things resembled many other colonia communities. The roads were dirt and sand with potholes large enough to tear a tire off. But it is not the holes and ruts that cause the real problems driving in Los Montes—rather it is, at times, the deceivingly deep pockets of sand that trap cars. The few stop signs that exist were usually ignored: stopping while driving in the sand is not a good idea. Driving in the sand is all about slow constant motion; or, if you are an adolescent male, fast, constant motion. The condition of the roads also limits the places you can actually drive, and it is important as a pedestrian to know where not to walk. There were no sidewalks, and often the best part of the road to walk on was also the best to drive on. There were the man-made ditches that often exist in colonias where pipes are being laid for some type of improvement. While avoiding the occasional oncoming car, you also had to avoid the much more consistent barrage of mean dogs. Most of the dogs in Los Montes were sweet old things that look like they had seen better days, but there were also quite a few that were violent and quite scary. You never know what a dog is like in a colonia, and it is always a good idea to give them a wide berth. Residents of colonias know where the mean dogs live and how far they will chase you. As hot as it was and as lazy as these dogs really were, if you ran when they start to bark, you could usually lose them in a few dozen feet.

One of the first sights a visitor to Montes saw was Doña Ida's store. Doña Ida was a real businesswoman. Not only did she own the larger of the two stores in the community, she was also a real estate developer and a poultry cultivator. Doña's lot was always full of chickens, tiny travel trailers, and truck shells set on tarps on the earth. The chickens came and went as they pleased, while the tiny trailers and camper shells sheltered migrant field labor that could not afford other accommodations. The other residents of Los Montes did not approve of Doña's renting camper shells and tents to recent immigrants. Their opinion was that she took advantage of these poor migrants by charging them way too much for nothing, but colonia residents also knew these field workers needed a place to live. Once past the miniature village of campers and truck shells, Doña's trailer came into view in all its splendor. Like most colonia homes, this was once a singlewide trailer, but had evolved into much more. Rooms were added on all sides and at all angles, at different heights and in different colors. One of these rooms served as the store where she sold soda, chips, candy, and other items.

Los Montes was declared a colonia in 1993, four years after its first occupants arrived. Flora and Esperanza, two of the leaders in this study, arrived in 1989 and were followed soon after by a number of other families. The colonia was approximately a half square mile. There were thirty households on slightly fewer than thirty lots. Several lots contained more than one household, often occupied by blood relatives. Los Montes was a very youthful community, with over 45 percent of the population less than fifteen years of age (Williams 2000). The vast majority of housing stock was made up of trailers with a few wood-frame homes. Many of these trailers were very old and in disrepair.

Valle de Vacas

Valle was officially designated a colonia in 1997 by Doña Ana County. It too is a HUD-approved colonia. Valle was home to approximately 250 families, living primarily in older trailers and a few stick-built homes. As the largest colonia in my sample, Valle presented problems for the study that the other colonias did not. There were only three COG-identified leaders in Valle and only a dozen or so other residents active in colonia politics through the COG at the time I was doing my research. While in New Mexico I interacted with approximately twenty Valle households; in

Montes or Recuerdos, this would mean I met nearly everyone in the colonia, but in Valle I only met roughly 10 percent of the households.

My first impression of Valle de Vacas came in the evening. It was still winter and the daylight ended abruptly just as people were leaving work for the day. I was accompanying Ernie and Mario to a community meeting, the first of many to get the organization in Valle back on its feet after an extended period of disorganization. Ernie had taken the job over from another organizer at the COG and the colonia, not yet used to the new organizer, had not gotten together to work on any projects. Valle is located on the interstate south of Las Cruces behind a series of dairies. The smell of over one thousand cows on a couple dozen acres is unbelievable, thus the name, Valle de Vacas, valley of the cows. This colonia always smelled like cows, and on a bad day—a really hot and windy day—the smell might make you sick. But as with most things, with time the smell was less obtrusive and only noticeable on really bad days.

The spatial isolation of colonias often means they are located in less-than-desirable areas near dairies, sewage treatment plants, or, worst of all, chicken farms. They experience environmental racism like many other marginalized groups who are subjected to unsafe and uncomfortable physical conditions because their lack of resources makes it difficult for them to organize and oppose the siting of environmental hazards such as dairies, chicken farms, and landfills (Cutter 1995; Pulido 1996). Several of the proposed sites for the new regional waste-water treatment plant were near colonias and the local Anglo land owners justified these choices with arguments that likened colonias to Mexico and discourses that equated Mexico with dirtiness and filth (Hill 2003; Vila 2000).

Structure of the Book

From this wide theoretical grounding, the book moves to the concrete ethnographical detail on which its argument rests. It is in the next four chapters and the epilogue that the daily practices of the women who lead the colonias will demonstrate the relationships between the activists, NGOs, and the neoliberal state that I have laid out in broad strokes here. Chapter 1, "The Production of Colonias as Neoliberal Spaces," describes and examines the particular historical and geographical conjuncture at which colonias developed in New Mexico. The chapter moves from the early history of northern New Mexico through the nativist sentiment of the early 1990s and argues that the historic reliance of the United States on cheap Mexican labor was an important factor contributing to the large Mexican immigrant population in the Southwest. When this historic reliance on Mexican labor intersected with the growing neoliberal retrenchment of the state, a large population of Mexican working poor was left without necessary affordable housing and social services. In typical neoliberal fashion, the private market was considered the best solution to this housing shortage, and the state, though lax subdivision laws and even more lax enforcement, allowed colonias to develop. As the problems colonias pose came to light, the neoliberal state, with the help of NGOs, began producing colonias as self-sufficient communities, creating a discourse that would solve their problems with a minimum of state intervention.

Chapter 2, "The Production of Women as Neoliberal Leaders," follows from the arguments presented in chapter 1 on the production of colonias as neoliberal Mexican working-poor communities and moves this argument on to the production of colonia women as neoliberal political subjects and leaders. This chapter focuses on the manner in which women are produced as leaders both by the state and NGOs. In particular, the focus here is on how women are recruited to leadership and the effects this has on both leaders and their colonias.

Chapter 3, "Empowerment and Politicization in the Colonias," focuses on the results of women's leadership, activism, and NGOs in the colonias in an effort to make connections between daily practices and the lack of politically progressive social change in the colonias. This chapter asks, what are colonia leaders accomplishing and in what ways are these accomplishments limited? This chapter argues that the desired goal of many NGOs that work in the colonias is the politicization of leaders and residents for the purpose of creating progressive social change. Yet, when closely examined, women's leadership produces only inconsistent empowerment of individuals and very little political awareness or sustained activism at the community level. This lack of politicization can be attributed in part to the individualistic nature of the neoliberal political subject as it is produced in the colonias by both the state and NGOs.

Chapter 4, "The Place of NGOs in Daily Life," continues this line of questioning and focuses more closely on how NGOs are implicated in the absence of progressive social change that confronts systems of power in the colonias. The growing importance of professionalism and competition among NGOs constitutes one of the primary obstacles to the development of politicization in the colonias. As NGOs take over more and more of what used to be state-provided social services, they are put in direct competition with each other for the limited state funding that is available. In order to be competitive and to gain funds, it is becoming more important for NGOs to present themselves as professional and expert organizations. In the colonias, this growing stress on professionalism, expertise, and market competitiveness draws NGOs away from the actual communities they aim to serve and focuses most of their efforts on administrative details. This distancing from the community can have devastating effects on the efficacy of NGO interventions.

The final chapter is an epilogue in which I document and discuss a return trip to the colonia of Los Montes seven years after the initial research took place. Through its focus on the state of leadership, activism, and community organizing in Los Montes, the epilogue revisits some of the key arguments and conclusions made in the prior chapters. It is here that I also propose ways the ideas and theories put forth in the book can be used by leaders and NGOs.

This book offers a close examination of the neoliberal state as it abandons working-poor communities in the U.S. Southwest and the responses that arise from these communities. In the chapters that follow, the development of colonias and their place in the border economy will be discussed as the foundation for a discussion of women's community leadership and its relationship to NGOs and the neoliberal state. My aim in this book is to contribute to and broaden the conversations on women's activism in the neoliberal era. It is my hope that in highlighting the linkages between women's activism, NGOs, and the neoliberal state in the colonias, this book can open new conversations and guide academics, activists, and NGOs in their struggles.

· I ·

Neoliberalism on the Border

CHAPTER 1

The Production of Colonias
as Neoliberal Spaces

El que quiere baile, que pague músico.
He who wants to dance should pay the musician.

I T WOULD BE WRONG to assume that neither state officials nor colonia residents view the existence of colonias as an offense to resident's basic rights as citizens. Rather, in many situations, the resource deprivation experienced in the colonias as an issue of basic rights is just not discussed. Yet, according to Orlando Cervantes, Chairman of the Doña Ana County Planning and Zoning Commission, "We have complete communities here without water, sewer or roads for access by emergency service. The government has failed in this case to meet its obligations to protect the people" (Nelson 1995). For Chairman Cervantes to believe "the government had failed to meet its obligations" by allowing communities to exist without basic services, he would have to believe it is within the role of the government to be sure people have these basic services in the first place. This is an assumption that is not obviously in line with the central tenets of neoliberalism: privatization and personal responsibility. Yet colonias are clearly neoliberal housing at its best. They are privately developed and brought from the lowest standards up to the most basic services by individual residents as a matter of personal responsibility, what could be more in keeping with neoliberal ideals? This chapter examines how colonias become one of the many spaces of neoliberalism.

I examine the history of the border region and the development of economic integration between the United States and Mexico and the neoliberal state's abandonment of immigrant communities in order to illustrate colonias as neoliberal spaces and this production of space as a site of convergence. It is at this site of convergence that I approach the book's greater question of how leadership is constructed and gendered across the shifting dynamics of neoliberal restructuring; nongovernmental organization

(NGO) interventions; transnational connections; and local, ethnic, and gendered relationships. Here, the foundation is set from which later discussions of the development of leadership, the role of NGOs, and the complex relationships between the two can be examined.

The Production of Colonias

Colonias are home to working poor on the U.S. side of the border, and a great many of the working poor here are Mexican immigrants. Over the years, the number of jobs available to the working poor in the region has increased rapidly. Economic development along this part of the border succeeded. Both agriculture and light production have increased in southern New Mexico and Texas (Larson 1995). But for all the economic development and job production, little has been done to create affordable housing for the workers attracted to the new jobs in the region.

One of the most enduring relationships along the border is that between Mexican labor, usually immigrant, and the U.S. economy. By tracing this relationship, I argue that colonias provide more than just homes for Mexican working poor on the border. They are an important site for the production of Mexican immigrants as self-reliant neoliberal citizens. A complex set of actors, structures of power, techniques of governmentality, and history combine in the production of colonias as impoverished communities in which the state is nowhere to be seen yet exerts its power in insidious ways.

This work attempts to question, pull apart, and complicate popular discourses on the colonias. Because these are rural, out-of-sight communities, many of the discourses surrounding the colonias are based on hearsay and assumptions and not on actual experience. Few politicians or journalists have actually visited colonias, but many pass judgment based on little more than the federal definition and the lack of infrastructure the definition implies. I examine and complicate the solely economic view, which holds that colonias exist because they pose the only logical alternative for housing the working poor (Ward 2000). Alongside the rather one-dimensional economic perspective, I highlight the disturbing political reasons why the existence of colonias fits well with the idea of neoliberal governmentality and an independent citizenry.

Colonias are complex and ambiguous communities. At first sight, they appear to be underdeveloped Third World slums. The dirt, dust, and garbage, combined with the run-down housing stock and unpaved roads, give these areas a seemingly deserted appearance at times. Much of the limited research that exists on colonias focuses on the following problems: their inefficiency as housing markets due to the lots' lack of appreciative value (Ward 2000; 2001); the environmental degradation by colonia residents through leaky septic tanks and cess pools (Carew 2001); and the health problems of colonia residents due to human waste contamination and generally unsanitary living conditions (Stevenson 2001). Yet my research shows that colonia residents remain optimistic about the future of their communities and have positive things to say about living in them. Clearly, colonias offer residents many of the aspects of daily life they desire. At the same time, colonia residents are not satisfied with the deprivation of resources that plague them daily. It is on this fine line between satisfaction and deprivation that colonia leaders must balance their activism. Women leaders in colonias exemplify the ambiguous and often contradictory situation in which they work. The greatest contradiction of women's activism in colonias is that if they do nothing, nothing will improve. But if they do something, they become the perfect working poor, as they are willing to create their own infrastructure and demand next to nothing from the state, not even what the state requires be provided for others.

The development of colonia communities neatly exemplifies the complexities of globalization. But we have to be wary of "top-down" or "impact" models of global economic restructuring and neoliberal policy because they alone cannot explain the development of the colonias. These models, which portray globalization as an all-encompassing and coherent force that easily over takes local cultures and meaning systems, must be questioned and engaged with critically. These theories can only go so far in explaining the growth of the conditions that led to the development of alternative housing on the border. They do not explain why colonias, rather than urban ghettos or working-class suburbs, became the most popular solution to this need. It is true that many colonias developed in a milieu of neoliberal retrenchment of the state and privatization of social services in the 1980s and 1990s. Yet these neoliberal cutbacks exist within a discriminatory housing market, especially on the border, and the inadequate and stigmatized public housing that does exist simply cannot house the growing population of Mexican working poor. It was in this particular convergence of limited housing, ethnic discrimination, neoliberal policy, and the prevalence of Mexican culture that colonias developed. It is critically important to always view the neoliberal political project in its case-specific forms. Like its frequent partner in crime, globalization, neo-liberalism manifests itself in very different ways in each locality, and it is this ability to morph itself and adapt that makes it such a powerful presence (Duggan 2003; Harvey 2005; Larner 2003).

The Production of Colonia Subjectivities

The colonia leaders this book chronicles are immigrants, farm workers, activists, mothers, wives, Mexicans, and Americans, sometimes all at once. The production of multiple subjectivities in the colonias can be examined as the result of a series of processes, geographies, and governmental relationships. The three central processes discussed in this project are global economic restructuring, political integration between the United States and Mexico, and immigration between the two countries. All three processes fit neatly under the greater heading of globalization. All three of these processes can also, in turn, be understood in the context of the ever-expanding specter of the neoliberal political project and its implications for the relationship between the state and both its citizens and noncitizens. The North American Free Trade Agreement (NAFTA), the growth of the border patrol, and the relationship that developed between Mexican President Vincente Fox and President George W. Bush are all phenomena that are tied to these processes. My focus in this chapter is on the increasing economic and cultural integration of the border region into the core of both countries. This discussion then moves on to develop an understanding of the production of colonias in two ways: First, I will look at the production of colonias as both a necessary and practical method to house Mexican labor. Second, I will look at the more subtle ways in which they produce immigrant labor as independent, neoliberal citizens whose actions can be taken as consent for their own infrastructure-lacking communities through the discourses of personal responsibility that permeate the solutions to the resource deprivation they experience.

The geographies involved in the production of the colonias all relate, in varying degrees, to their location on the border and the historical importance of the border region to the development of both the United States and Mexico. The border region is often referred to as the area one hundred kilometers north and south of the international boundary, and, in 2000, this region contained a population of 10.6 million in both Mexico and the United States (U.S. Environmental Protection Agency 2000). The border region has always had a troubled relationship with the seats of power in both nations. Its distance from Mexico City and Washington, D.C., led to a sense of isolation and independence that made the border especially attractive for those seeking privacy. This distance from the interior also led to the development of strong cross-border ties, as many border dwellers on both sides felt that they had more in common with their neighbors on the other side than they did with their fellow citizens thousands of miles away in the interior of their home country. Daily life for many on the border provides a binational experience. The bulk of the fieldwork for this book was done pre-September 11, when this binational experience was for many a much more seamless experience. Colonia communities exist almost entirely outside city limits and on the border in some of this country's poorest counties. Isolation is a theme of daily life and an important way in which colonias are produced as Other, foreign, and necessarily hidden. It also makes it slightly easier for overburdened counties with insufficient tax bases to treat colonias with an "out of sight, out of mind" policy whenever it is convenient. Yet colonias were intentionally located in rural areas, where the farm workers can provide labor for the growers easily. The informal nature and lack of infrastructure in the colonias add to their production as marginal, incomplete, and underdeveloped. This geography creates communities that are isolated and circumscribed from other local, more urban communities yet are highly integrated into the regional agricultural economy and connected to their transnational roots.

The relationships that most often define the production of the colonias are those that place the colonias within the processes of globalization. The mutual dependence of American employers and Mexican labor shapes economic and political integration between the United States and Mexico. Over time, constant immigration flows for labor and economic opportunity from Mexico to the border region of the United States has created transnational communities. Though both the Mexican and U.S. governments understand this transnational relationship, they are not equally willing to regularize these flows. While Mexico takes strong steps toward supporting many of the undocumented flows and aiding the population of Mexican working poor in the United States, the United States moves farther away from notions of aid for the poor and closer to the growing neoliberal ideology that dictates a minimal state role in social reproduction and the circumstances of daily life. It is this relationship between the neoliberal state and the colonias that sheds the most light on the production of colonias and their women leaders. In the colonias, the neoliberal political project has found the ideal solution to the ongoing problem of affordable housing for the working poor: privatized and individually generated housing at no up-front cost to the state.

Here I draw on Aihwa Ong's work on neoliberalism. Ong starts from the premise that neoliberal technologies of governance can be traced back to a "biopolitical mode of governing that centers on the capacity and potential of individuals and populations as living resources that may be harnessed and managed by governing regimes" (2006, 6). As a purposefully produced space in which to house cheap and readily available labor, colonias fit well with the concept of a "biopolitical" mode of governing. In the colonias, what Ong calls "technologies of subjection" are at work informing "political strategies that differently regulate populations for optimal productivity, increasingly through spatial practices that engage market forces" (6). Colonias are clearly "spatial practices that engage market forces," as they represent the privatization of affordable housing. The reason colonias exist is simply to house cheap labor near where it is needed, clearly fulfilling Ong's criteria of "regulating populations for optimal productivity."

The production of colonias is a complex and multiply determined process that involves a number of historical circumstances and geographic relationships. No single reason led to the creation of the colonias. Together, however, these processes represent a particular historical and geographical juncture at which colonias developed in order to provide a place to house Mexican working poor along the border. Historically, Mexican nationals have crossed the border, both with and without the necessary documentation, to provide the labor that runs the region's agricultural economy. The government briefly legalized this stream of cheap labor during World War II through the Bracero program, and the first colonias in Texas were created during the Bracero program in the 1950s to house Mexican farm labor (Donelson 2004). When the Bracero program ended in the mid-1960s, the Border Industrialization Program (BIP) marked the introduction of the border region into the processes of global economic restructuring. The rapid growth of the border region, thanks to global integration and NAFTA, led many Mexicans from the interior of Mexico to the border in search of employment. Only a short distance connected Mexicans to better employment opportunities, and many made this move. The growth of the border region and the increased numbers of Mexican immigrants in the United States led to an anti-immigrant backlash in many border states. The Immigration and Reform Control Act (IRCA) in 1986 was one of the first official responses to the growing issue of Mexican immigration. IRCA provided amnesty and documentation to many Mexicans in the United States and allowed them to reunite their families. But, once united, these families needed housing. The lack of affordable housing on the border hindered their resettlement and reunification. In an attempt to take advantage of this new group of legal residents searching for housing, developers in the Southwest used the lax subdivision laws in New Mexico to their advantage and created colonias without even the most basic infrastructure. However, even though the housing lots lacked basic infrastructure, they had many characteristics that immigrants desired, such as a culturally Mexican community and a tranquil, rural environment in which to raise a family.

Historical Background

The story of the development of colonias along the U.S–Mexico border is essentially the story of Mexican immigration to the United States. Because the southwestern portion of the United States used to be Mexican territory, the history of Mexican immigration to the United States differs from that of other early immigrant groups.[1] Mexicans have the longest history of immigration to the United States. Indigenous people from what is now Mexico were an important source of labor in the region as early as the sixteenth century, when the Spanish used them alongside Native Americans to build missions and settlements. Since that time, Mexicans have continued to supply the majority of manual and unskilled labor along the border. Prior to the annexation of Texas and New Mexico in 1848, Mexicans provided the labor the Anglo settlers needed to homestead and develop the region.

Labor demand and supply primarily dictated the history of immigration between the two countries, and this pattern continues today. Geographical proximity led both to the critical integration of Mexican immigrants into the United States' agricultural economy and to the competition for markets that had devastating effects on the Mexican economy. The late twentieth century was a very hard period for the Mexican economy. The structural adjustment policies of the 1980s left Mexico, like much of Latin America, with even more poverty and a more polarized distribution of wealth than it had before (Skidmore and Smith 2001). The growing economic integration between the United States and Mexico and the flood of heavily subsidized U.S. agricultural products into Mexico after NAFTA's implementation in 1994 exacerbated this poverty. Major agricultural subsidies in the United States meant that American corn could be sold in Mexico for less than production cost. Between the artificially low prices of American farm products and the growth of contract farming and large-scale agribusiness in the United States, Mexican farmers could not compete. These unemployed rural farmers were forced to migrate for work. Many headed toward the border and then into the United States.

Early Migration History: The First Mexican Immigrants

Mexican immigration to the United States started in earnest after 1848, when the southwest region of the United States ceased to be Mexico and the Mexican inhabitants of the area either became American citizens or returned to Mexico. During the rest of the nineteenth century, Mexican immigration to the United States went primarily unquestioned. The border growth boom began during the presidency of Porfirio Díaz (1876–1910; Lorey 1999, 35; Massey 2002). Díaz created a political stability that Mexico had not seen since its independence in 1821. His motto was "Order and Progress," and the economic growth and stability that occurred during the Porfiriato encouraged American capitalists to invest in the border region. With dramatic changes occurring in both countries, the late nineteenth and early twentieth centuries experienced mutual growth along the border. This area was scarcely populated at the time, and any additional labor was generally accepted. Employers on both sides of the border boom welcomed Mexican labor.

Today many in the United States view Mexican immigration as a major problem that must be controlled through restriction. But it was the Chinese, not the Mexicans, who were the focus of the first restrictionist immigration law in 1882. Between the late 1880s and the early 1900s, immigration control on the southern border focused on stopping the Chinese who tried to avoid the 1882 Chinese Exclusion Act by entering the United States through Mexico. Until 1917, immigration control on the southern border was fairly relaxed for Mexicans, who passed across the border with little or no trouble.

Mexicans remained, in many ways, hidden in the Southwest for years, while national attention focused on the growing numbers of southeastern European immigrants entering the country. The tradition of keeping Mexican labor "out of sight" has a long history along the border; the colonias were not the first form of housing for Mexican immigrants that that sought to position these communities in out-of-the-way areas (Ward 1999). The development of these immigrant communities in secluded areas allowed the population of Mexican immigrants to grow to levels that might otherwise have alarmed Anglos if they had been more visible. These hidden geographies of immigrant communities are central to the disciplining and production of immigrant subjects as citizens in the United States and shape the later treatment and location of immigrant communities.

Earlier immigrants of Northern European background who controlled the nation's immigration policy at the time considered Southern and Eastern European immigrants unwanted. Growing restrictionist and nativist movements shaped immigration policy at the turn of the century. The popularity of scientific racism during this period led to strong sentiments regarding the "wrong type" of immigrants, who were usually those of Mediterranean or Asian descent. Fear of the "wrong kind" of immigrant led the country to pass the 1924 National Origins Act (NOS), which severely limited the numbers of immigrants from "less desirable" areas (Gjerde 1998). An unintended side effect of NOS was to draw attention to the growing numbers of Mexican immigrants in the United States. With the number of immigrants from Eastern and Southern Europe on the decline, the number of Mexican immigrants became a much higher percentage of the total immigrant population. This growing visibility alerted the nativist and restrictionist elements, who began the campaign to limit Mexican migration to the United States. If Mexicans were to enter freely, then all the work the restrictionists had done to protect the racial purity of the United States from non–Northern European blood would be negated (Reisler 1996, 30). Restrictions had to be applied to Mexicans as well.

Farmers in the Southwest did not agree with the restrictions on Mexican immigration. They first voiced their dissent in 1917, when the United States passed an immigration act that established several tests migrants had to pass and a head tax they had to pay before they could cross the border (Craig 1971, 7). Growers argued that to limit or stop Mexican immigration would signal the end of the border boom. Without Mexican labor, the growth of the border region's agricultural industry would be destroyed.[2]

Growers also argued that Mexican laborers did not take jobs away from American workers in the way the restrictionists believed European immigrants did. They argued that the jobs that Mexicans took were the ones Americans neither wanted nor were suited for. This argument, based on ethnic and racial differences and labor demand, explains the early placement of Mexican immigrants in unskilled and often-unwanted jobs, a trend that continues today. Finally, the growers pointed to the temporary nature of much Mexican labor. Mexicans came from nearby and could return to their homeland when their labor was no longer needed. Repatriation policies were implemented during economic downturns throughout the twentieth century (Garcia y Griego 1996, 49; Reisler 1996, 37). Between 1953 and 1955, "Operation Wetback"—the most famous example of repatriation—deported two million Mexicans, as well as many U.S. citizens of Mexican descent simply based on their "Mexican" appearance (Lorey 1999, 121).

World War II and the Border's Growing Agricultural Economy

With the onset of World War II, the border economy entered an even greater boom period. With abundant electricity and water from projects such as the Hoover Dam, the border states attracted many of the new wartime industries. The new industries required even more labor, especially when American men left to fight overseas. American women filled many of the jobs left vacant by departing soldiers, but in the border's agricultural industries women often were considered an inappropriate substitute. Mexican men were the replacement of choice. This period also signaled a tremendous increase in industry and population in the Mexican border states. Collaboration between the United States and Mexico along the border increased the supply of raw materials Mexico extracted for the fabrication of munitions and other wartime industries in the United States (Lorey 1999, 89). The increased production on the border in Mexico meant more and more people moved to the border region from the Mexican interior. The increased population in Mexican border states provided the United States with an even greater supply of cheap, temporary, Mexican labor.

During the war, the U.S. government, at the prompting of growers in the Southwest, decided to authorize the migration of Mexicans as farm labor. The Bracero program began in 1942. The program, named for the Spanish word for "arm," was essentially just that: a program to supply the region with the extra arms it needed to continue farming (Chavez 2001, 204). During the twenty-two-year course of the Bracero program, approximately four and a half million Mexican nationals crossed the border under its auspices (Craig 1971, ix; Massey 2002). The debates surrounding the controversial Bracero program still plague immigration discourse today.[3] In his book, *The Bracero Program: Interest Groups and Foreign Policy*, Richard Craig discusses both sides of the debate (1971). As with earlier discourses surrounding Mexican immigration, those that supported the Bracero program were primarily growers and others who needed cheap labor. The opposition comprised a more diverse group. One of the main groups that fought against the program was organized labor in the American Federation of Labor and Congress of Industrial Organizations and its member organizations. Organized labor believed the Bracero program and its masses of nonunion labor would impair their efforts to unionize the nation. Human rights organizations also protested against the Bracero program because they feared that Mexican workers would not be treated fairly.

Between 1942 and 1951, the program operated under various government authorities. It was not until 1951, with the passing of Public Law 78, that the Bracero program had a single statutory basis (4). Public Law 78 initiated the circumstances under which Mexican temporary labor could be recruited. The use of *braceros*, or Mexican laborers, was restricted to areas where there was a bona fide shortage of workers, and only if braceros did not lower wages for natives. In addition, the fair treatment of Mexican nationals in the United States was guaranteed. The guarantees included wages at least equivalent to those paid to native labor for the same task, employment for three-fourths of the contract period, adequate and sanitary housing, occupational insurance at the growers' expense, and free transportation back to Mexico at the end of the contract period (4). In his discussion of these guarantees, Craig points to the very vague nature of the wording used. Growers interpreted words such as "adequate and equivalent" in many ways, and most often the braceros were adversely affected by these interpretations. This was yet another example of the disposable nature of Mexican immigrant labor in the United States. For many braceros and their families, a lifetime of "inadequate" housing began while living in farm-worker camps scattered across the United States. Many existing colonias have the look and feel of Depression-era Dust Bowl migrant farm-workers camps. It is not much of a stretch to imagine developers coming up with the idea to create these illegal subdivisions after driving by or through existing farm-labor housing.

The Bracero program shaped the lives of many on the border, and quite a few current residents in the colonias are there because their families moved to the border region of Mexico when a father or grandfather joined the program. In my research, I met several leaders and activists whose first experiences of the United States came in the form of stories their male relatives told of their adventures as braceros. Estella recalled, "My papa was a bracero. Always, since we began to grow up, we always saw he was never in the house. Because he came with the braceros, he always worked here in the United States."[4] In order to make working in the United States less complicated for their families, many braceros relocated their families to the border so they could more easily return home after rotations working on the northern side of the border. Once on the border, work in the United States was only a few steps away or "across the bridge." Estella continues: "He crossed the bridge and there were buses that took the people to different parts to work . . . He was in Chicago, he was in Wisconsin, in Colorado, in various places."[5] The familiarity Estella's family developed with the United States, and particularly the U.S. borderlands through her father's work as a bracero, made it easier for her family to make the move across when her father decided it was time.

In the period since the Bracero program ended, Mexicans have remained the preferred labor in agribusiness along the border. This is due, in part, to the economic strains Mexico has experienced in the period since 1964 and, in particular, to the explosive growth of the Mexican population and the lack of similar growth in jobs, which pushed many into the United States in search of employment.

Mexican agricultural workers on the border still live with the legacy of the Bracero program. Because there is always a surplus of Mexican labor, both in the United States and just on the other side of the border, U.S. growers consider Mexican workers temporary labor that can be paid the lowest wages. Currently, the low levels of unionization are considered by border scholars to be a direct result of discourses of disposable Mexican farm labor that developed during the Bracero period.

The Current Phase of Globalization
and the Growth of the Border Region

When the Bracero program ended in 1964, the Mexican government feared that the decline in legal employment for Mexicans in the United States would have devastating effects on the Mexican economy. The Mexican government created the BIP to avoid such problems. The BIP was established "with the twin goals of stimulating the manufacturing sector of the depressed economies in the northern states and providing employment for workers displaced by the end of the Bracero program in 1964. The BIP was also established to respond to changes in the world economy" (Lorey 1999, 104). These changes included the development of communication and manufacturing technology that facilitated the beginning of export-processing zone production. With its geographical proximity to the United States and large unemployed population, northern Mexico proved an ideal location for U.S. manufacturing. The BIP lead to the creation of the *maquiladoras*, or export-processing plants, that are now well known all along the border. *Maquilas*, as they are commonly known, import raw materials and components duty-free from the United States and assemble them before they export the finished product back to the United States. Proximity is not the only benefit maquilas offer. Because of their location in the free-trade zone between the United States and Mexico, export processing in maquilas also offers a tax break. This means products imported from maquilas to the United States are taxed primarily on the "value added," or the amount of labor put into them. Due to the low wages maquila workers are paid, this is a relatively small amount. Maquilas proved very quickly to be an asset for both U.S. and foreign producers, and they expanded at an astonishing rate. In 1965 there were 12 maquilas; by 1998, this number had grown to 2,978. [6]

NAFTA represents a pivotal point in the history of economic integration along the border. The Mexican government saw NAFTA as a way out of the crisis of unemployment that had hit the country. But along with NAFTA came a very serious currency crisis in which the Mexican peso's value plummeted. The peso crisis led to an even greater wage differential between Mexico and the United States, and the buying power of Mexican workers sank lower and lower. NAFTA fixed very little. Rather, it created an even greater motivation for Mexican labor to cross into the United States for work. NAFTA did more than just move jobs south of the border and facilitate a peso crisis; it also gave employers in the United States the perfect threat to hold over domestic labor. U.S. employers use the threat of moving more jobs to Mexico or China, where wages are much lower, to suppress wages and keep them more "competitive." They tell workers they cannot make a profit if they must pay higher wages due to health and safety regulations or union bargaining (Petras and Veltmeyer 2001). NAFTA proved to be a sign of the contradictions that characterize the immigration policy of the United States: at the same time we opened the border to trade and encouraged the cross-border flow of money and goods, we cracked down on the flows of people (Massey 2002).

The connections between globalization on the border, the transnational economy of the region, and the immigration of Mexican nationals to the United States are many. In an effort to understand these ties, I want to examine how the globalization of the U.S. economy and politics has directly affected immigration flows to the United States. The importance of social and economic relationships between countries in the development of migration streams is now commonly acknowledged (Sassen 1996; Staudt 1998). For the United States, this contact takes place through its central role in the growth of the global economy. In the case of Mexican immigrants to the United States, familial, political, economic, and cultural connections between the two countries lead to increased migration from Mexico. Through economic and technological investment, the expansion of U.S. media outlets, and political interventions, the United States has developed a large-scale presence in Mexico. This presence creates a familiarity with the United States that makes immigration to the United States that much easier. Even more important are the kin and social networks that decades of Mexican immigration to the United States have produced. Small Mexican villages have unofficial "sister" cities across the United States where large numbers of villagers live for parts of the year and make money to spend in Mexico during their frequent visits home (Nichols 2002). When examined in this way, it becomes clear that immigration is not solely domestic or economic. The ties between globalization and immigration make it necessary to view immigration as a set of multiple scaled and interrelated processes (Chavez 2001; Sassen 1996).

If we view immigration as a set of processes played out at scales ranging from that of the body to the state, we must then situate colonias, as immigrant communities, within both local and global processes. Building on my earlier arguments surrounding economic integration on the border, the growth of export processing in Mexico demonstrates how the border economy has grown as a transnational unit. As maquiladora production expanded on the Mexican side, corporate offices and more technical production appeared on both sides. Mexican migration to the borderlands has a long history and was in full effect before the BIP began in 1964, yet the BIP created an even greater draw for Mexicans from the interior of the country (Sassen 1996). Once migrants arrive on the border, many cross over into the United States. Some work in maquiladoras for a few years first, and others, especially men who often cannot find maquila work, cross over immediately (Lorey 1999).

Maquiladora production, although originally envisioned by the BIP to employ men left unemployed by the termination of the Bracero program, soon proved to be one of many feminized sectors of the global assembly economy. In an era when the use of women's labor was growing in popularity, maquiladoras match this move toward assembly work based on the use of women's docile, agile, and adept labor, which was perfect for the intricacies of assembly (Elson and Pearson 1981). Yet, as time goes by, the numbers of women in maquila production is evening out, as men become as popular a source of labor as women (Tiano 2006). This may be because of the increased variety of goods being produced or because there are simply more men available, and transnational corporations are now choosing to employ men as well (Tiano 2006).

All but one of the women leaders described here were born in Mexico and migrated to the United States in their teens. Some, like Estella, came with their parents as a family unit, often after their fathers had crossed the border for work under the auspices of the Bracero program. Others came on their own in search of work, excitement, or romance. Several women were born in northern Mexico, but not on the border, and they moved with their families to take advantage of the border's expanding economy where they found work in the maquiladoras. Whatever the reason that brought these women to the United States, they have stayed here, have become documented residents or citizens, and have made themselves a home in the colonias. Five of the women in my core sample came to the United States by themselves for work. These women first moved close to the border with their families and then made the move into the United States on their own or with the help of a relative or family friend.

Marie, who moved with her mother, father, and her eleven siblings, was born away from the border and moved to Juárez with her family to look for work opportunities. Her father found work on the Mexican side, but she thought there was more opportunity in the United States. Marie, like many of the women I met, made her way into the United States with the help of a "friend." This woman was really more of an acquaintance, and later their relationship was that of a mentor and apprentice. But this was not the relationship that mattered. The job was less important to Marie than the trip across the river and the insight into how to cross and how to make it once on *el otro lado* (the other side). As Marie describes the experience, "I was eighteen years old. Then a woman we knew, who lived there in Juárez and had gone to live here [United States], she came to bring my sister, my youngest sister. She brought her to work here with her because they had worked together in Juárez." Once Marie saw her sister's success, she made it clear to this woman that she, too, wanted work in the United States: "And when she knew that I wanted to work in another place, she brought me to work here as well."[7] Marie worked with this woman for three months cleaning houses around El Paso and then left when the woman could no longer pay her. But during this time, Marie met the man she would later marry. He was also from Mexico but had been in the United States since he was a child.

Like many women immigrants to the United States from Mexico, women in my study heard stories about better work opportunities and wages in the United States from friends and relatives who had worked on the other side in the past. As is common along the border, in several cases the women crossed over in the care of older women who had established jobs cleaning houses or caring for children and needed a helper or "apprentice." Apprenticeship is very common for young Mexican women when they first enter the United States (Hondagneu-Sotelo 2001). Although the relationship between the apprentice and their mentor can often be exploitative, women who have mentors are also more likely to have a place to live and enough work to support themselves. This was not the case for Flora. Flora crossed the river from Juárez in her best clothing, which then became her only clothing. She had no job and no place to live. She spent the next couple of days trying to find something else to wear so she could wash the clothes in which she had crossed the river. Flora now speaks fondly of that outfit and how she wore it for weeks because it was all the clothing she had.

Due to the decline and reorganization of manufacturing, more low-wage, service-sector jobs exist in the United States. Flora and others are drawn to these jobs because they offer better pay than those they can get in Mexico. There has also been a rapid increase in the casualization of work. Immigrants, both documented and undocumented, often fill these new and less-desirable jobs that many long-term U.S. residents do not because these jobs have no benefits, pay very low wages, and are often only part time. These changes in production in the United States directly affect immigration flows in the form of "pulls" for immigrants.

Here the process of immigration comes full circle. The United States invests in Mexico, and this investment leads to a growth in urban factory production at the cost of the rural agriculture sector and, in turn, helps to create a large unemployed group of formerly rural agricultural workers. At the same time, the nature of production is changing in the United States, and many manufacturing jobs are moving overseas to take advantage of cheaper labor.

Immigration Becomes a "Problem"

Although colonias existed prior to the 1980s, there was a burst of colonia development around the mid-1980s. Manuel was the first person to explain this timing in any detail for me. Manuel had been a farm-worker organizer in the 1980s. When I met him, he worked in the local Catholic diocese's social-justice office. Like quite a few Mexican American radicals of his generation, he has found an interesting balance between his Marxist leanings and his place in the Catholic Church. Manuel explained to me that many colonias were developed in response to the IRCA in 1986.

I would later revisit this explanation for the rapid development of colonias in the 1980s in an interview with Mario, one of the community-organizing group's main organizers. Mario organizes Recuerdos and several other colonias near Hatch; he is also Estella's brother. Mario helped me when I wanted to get out into the fields and work alongside colonia residents to get an idea of what agricultural work was all about. He set me up with a local labor contractor and made sure I was on the same crew as Juana, one of the few colonia leaders who still worked regularly in the fields. While I was out with Juana, I noticed that there were quite a few couples and even families that picked together, some with teenagers or young-adult children who joined them, and others whose small children waited in their vehicles playing while they were working in the fields. When Mario mentioned that prior to 1986 the fields were a totally different place, I immediately thought of my experiences in the onion field. I asked if, prior to 1986, my experience would have been different. Mario responded that, yes, Juana, or any other women, would not have been there. It would have been all men who "also lived on the ranches or farms and the farmers provided housing, but only a place to stay and sleep."[8] Mario explained, that once families came over these men moved with their families to the colonias.

In support of my argument that Mexican labor in the United States is viewed as temporary and disposable, this section examines how the United States deals with these flows domestically, emphasizing that, historically, immigration to the United States was selectively enforced. When labor was in demand, the border was more porous, allowing more prospective workers to cross (Staudt 1998, 37). Prior to the production of the colonias as communities of independent neoliberal citizens, the state dismissed its responsibility to Mexican immigrant labor by repatriating excess Mexican workers. The porosity of the border may be changed by means of policy changes, by issuing more work passes, by cutting the number of border patrol agents in a particular area, or by simply turning a blind eye to illegal crossings (Dunn 1996). When unemployment increased in the United States, the border was tightened by a more vigilant border patrol and became increasingly more impenetrable. The selective enforcement of border control based on the labor needs of the region's industries has a long history that extends from the first border boom to the present. While Mexicans who live on the border in Mexico and commute to the United States for employment are all too aware of the fluctuating porosity of the border, many Americans hope and believe the growing border patrol presence and the development of the U.S. Department of Homeland Security and its stress on decreasing undocumented immigration will make the border more impenetrable.

The late twentieth and early twenty-first centuries have seen the rebirth of a very strong nativist, anti-immigrant sentiment in the United States. Drawing on many of the same arguments used during the country's first nativist period in the late 1800s and early 1900s, the nation's current obsession with "illegal" immigrants is both controversial and contradictory and heightened by September 11th. The arguments used by nativists are just as racist today as they were in the nineteenth century. The selective use of these arguments in border states against those of Latin American descent continues as well. The states of California, Arizona, Texas, and New Mexico, among a growing group of others, depend on the labor of Mexican immigrants, many of whom are or have been undocumented: "Both U.S.

citizens of Mexican descent and undocumented Mexican immigrants played a major part in the postwar economic boom on the border, particularly in the growth of manufacturing and service sectors" (Lorey 1999, 101). But this important contribution by Mexican immigrants did not stop the majority of the population along the border from creating and supporting a growing discourse on the "immigration problem." The culmination of this discourse is clear in California's Proposition 187, which passed in 1994 (Calavita 1996; Wright and Ellis 2000, 207). Proposition 187 was called the "Save our State Initiative" and was designed to severely limit the rights of and services provided to undocumented immigrants, including their children, working and living in California. But this did not happen as the federal courts tied it up by debating its constitutionality, before the state's next Democratic governor, Gray Davis, permanently stopped the proposition (Nevins 2002).

Based on the central importance of immigration to the production of the colonias, IRCA can be seen as a key point in U.S. immigration policy as Mario and Manuel explained to me (Dunn 1996; Nevins 2002). IRCA was based on the idea that once long-term undocumented workers were documented, the state could crack down on the border in order to stop new, undocumented immigration and to go after recent undocumented workers more easily. IRCA offered amnesty and the chance to legalize one's status in the United States under two main categories. The first individuals allowed to apply for legal status in the United States were those who could prove they had been in the United States for at least eight years. The process of obtaining amnesty was very long and complicated, usually requiring trained legal assistance. Agricultural field workers, including many colonia residents, used the second form of amnesty more often. This process required proof that one had worked in the United States at least ninety days in the last year. Most farm workers could prove this with pay stubs.

Employer sanctions were meant to be a central part of IRCA. Employers caught with undocumented workers on their payroll would be fined. Yet fines are rarely applied, and for the most part, employers use U.S. Immigration and Naturalization Service (INS) inspections to get rid of undocumented workers when they no longer need them.[9] Selective INS raids are currently one of the most popular forms of repatriation. Employers not willing to risk INS sanctions commonly use subcontractors in the hiring of labor. Rather than hire their workers directly, employers go through a labor contractor who is responsible for certifying the legal status of the workers. Then, if an INS raid turns up undocumented workers, it is the contractor, not the factory or ranch owner, who is held responsible. Just as border enforcement has been historically determined by labor demand, so, too, has the enforcement of employer sanctions.

In keeping with the central tenets of the neoliberal political project, border states such as New Mexico, Texas, and California made little effort to help settle the recently documented immigrants IRCA created. The growing "immigration problem," as many viewed the increasing numbers of immigrants, especially visible in border states, combined with the constant demand for cheap and readily available labor, exerted enough pressure to lead to the creation and implementation of IRCA but not enough to hold the state responsible for the integration of the newly documented U.S. residents into local communities. Following the tradition of expendable Mexican labor, neither local and federal government nor growers took any responsibility in the provision of affordable housing for the newly legal immigrants created by IRCA. After IRCA, the primarily male undocumented farm workers who used to secretly live on the ranches where they worked were free to move about as they pleased. More importantly, many were able to bring their families over to join them. The arrival of wives and children created a dramatic change in the dynamics on many farms. Farmers who had previously housed their laborers in tents and shacks no longer felt comfortable doing so with women and children present (Paster 1993, 7). According to Mario, "People moved to colonias because if they brought the family they couldn't live at the farm anymore because they had the children or the daughters. Many people do not allow the daughters or the sons to be living inside the [farms], only the men."[10] Farmers were not comfortable with children living on their back lots; it was a level of responsibility they were not willing to accept most of the time. Suddenly, the border region was filled with recently legalized families in search of housing. The increased permanent settlement of workers who had previously been circular migrants, or those crossing without documentation to work the summer agricultural season and then return to Mexico for the winter, was an unintended consequence of IRCA. Once IRCA passed, the border became more secure and harder to cross, forcing many undocumented workers to settle and not risk returning to Mexico. A serious lack of affordable rental housing still plagues the border as it does much of the United States (Massey 2002). For the newly legalized farm workers and their families of the mid-1980s, colonia lots offered one of the only available alternatives to renting.[11]

The relationship between the state and colonia developers is rather opaque. Colonias are illegal settlements, yet they fulfill several essential neoliberal mandates and were in keeping with the privatization that was washing across the United States during the 1980s and 1990s (George 1999). Colonias provide a privatized solution to the lack of affordable housing on the border. Privatized solutions were preferred between the mid-1980s and mid-1990s when colonias grew most rapidly in New Mexico. According to Lisa Duggan, "When the state acts in the 'public' interest—providing housing for the poor or protection for the environment—that can be intrusive, coercive, and bad" (2003, 13). Colonias are popular not only because they are a private rather than public form of affordable housing but also because, through their resource deprivation, they tend to breed citizens who appeal to the neoliberal political project because they display great levels of personal responsibility. When faced with life in a colonia with no electricity, running water, or paved roads, colonia residents take the initiative to acquire these resources on their own, often with little or no state intervention. If the neoliberal political project wants to have less to do with the social reproduction of its citizens, the colonias are a step in that direction, even if they are illegally developed.

An important response to the "immigration problem" that led to the creation of colonia communities was the militarization of the border. Throughout the early 1990s, INS increased security along the nation's southern border, pushing undocumented crossers out of urban areas and into the desolate and dangerous mountains and deserts (Dunn 1996). The INS called this process of increasing the numbers of border patrol agents and vehicles "Operation Hold the Line" (Massey 2002; Nevins 2002). In her book, *Free Trade?: Informal Economies at the U.S.-Mexico Border*, Kathleen Staudt draws a connection between the timing of "Operation Hold the Line" and the beginning of the NAFTA era. "Operation Hold the Line" began, according to Staudt, on September 19, 1993, "the seasonal eve of NAFTA's New Year transition" (Staudt 1998, 162). The argument behind the militarization of the border was simple: if the jobs were going to go to Mexico, the workers needed to stay there. There would no longer be a great need for Mexicans to cross into the United States for jobs. The blockade also pleased the growing numbers of anti-immigrant nativists along the border. The militarization of the border in the early 1990s changed the lives of all residents on the border (Nevins 2002; Staudt 1998).

With the increased "security" on the border, crossing for employment became harder for many undocumented workers. Undocumented workers found fewer and fewer places to cross the border without detection, as the blockade forced them into the deserts and mountains where crossing took longer and posed more danger (Nevins 2002). Documented workers were affected as well, especially through their relations with undocumented family members and friends. Most colonia residents are fully documented and have little reason to sneak past border patrol agents. Still, nearly every colonia family I met had at least one relative that visited regularly who lacked documentation. This situation means that most colonia residents are affected by the increased security on the border.

Angie's story illustrates the daily presence and threat the border patrol poses for many colonia residents. During my research, I met several colonia women who worked part time for local NGOs. NGOs sought out these women because they were the best way to get access to the colonias, and the women usually welcomed the extra money. Angie started as one of these women, but when I met her she was in charge of a local parent's resource center. Angie is a citizen of the United States with a Social Security card and all the other benefits of citizenship. But this does not mean her life is free from encounters with the border patrol. On the way back from a routine shopping trip to Las Cruces, Angie had a run-in with the border patrol that has had lasting effects on her reputation. Angie is a good-hearted person who goes out of her way to help others. When she noticed three young men on the side of the highway a mile or so past the border patrol checkpoint, she offered them a ride in the back of her pickup. They enthusiastically accepted her invitation in Spanish and hopped into the truck. Angie barely had time to pull back out onto the highway when a border patrol vehicle came speeding up behind her with its lights flashing. A confused and anxious Angie pulled over immediately. Angie later told people that she could not figure out what was happening. She had passed the checkpoint with no problems, and the men she picked up had to have documents or they would never have made it through the checkpoint themselves. If they lacked documents, they would have been crazy to walk along the highway in plain sight. Angie soon found out the men were not crazy; they were naïve, and the border patrol was playing a game of cat and mouse with them. The border patrol had been tracking the men ever since they first spotted them trying to get around the checkpoint through the desert. When Angie stopped to pick them up, they thought she was an accomplice. After she was taken into custody and questioned for several hours, she was released. The men were jailed and later deported. But for Angie, the damage was done. In a small, conservative town like Hatch, rumors have a way of serving as the truth. After this incident, Angie feared for her job and her reputation.

Since the early 1990s and the advent of "Operation Hold the Line," documented and undocumented workers alike have found that increased questioning and harassment at border crossings made their crossings more difficult. Even Mexican Americans found that they had to answer more questions at checkpoints and that the increased numbers of border patrol agents meant a greater likelihood that they would be stopped on the street in certain neighborhoods simply for looking "Mexican." With the increased difficulty of crossing the border daily for employment, many Mexican workers decided to stay in the United States and settle rather than migrate. Once they decided to settle, these workers experienced firsthand the lack of affordable housing on the border. Local developers quickly saw this gap in the housing market. They viewed the growing number of Mexican farm workers seeking land to buy as the perfect way to unload unusable farmland at a high price. This led to the development of many new colonias along the border.

Legislation and Colonias

Colonias developed in the El Paso–Doña Ana County region in what Larson calls a "regulatory vacuum," or what I would describe as a neoliberal environment (Larson 1995, 97). In a jurisdiction with a standard system of regulation, a developer must provide access to public services, such as wastewater and electricity, and a minimum level of infrastructure, such as roads, in order to get approval for the subdivision. In Texas and New Mexico, between the 1970s and 1990s, such regulations did not exist. Why was there so little regulation? After all, if there had been more regulation, thousands of colonia families would have been protected from the very difficult living conditions in which they now live. There are, of course, many relationships that contribute to the historical lack of regulation along the border in Texas and New Mexico, but one in particular has had lasting effects on colonia communities.

This is the "don't legislate me" attitude that many border residents exhibit in their relationship to the state. Self-reliance and independence are the hallmarks of the U.S. West (203). New Mexico is well known as a favorite spot for Old West outlaws to call home.[12] The open frontier and limited population centers made it easy to "hide out." To this day, many people still move to New Mexico because of the privacy offered by this sparsely populated state. I argue that New Mexicans, and border residents in general, seek this privacy because they want to do with their land as they see fit, without being regulated by the government. The state of New Mexico appears to have acknowledged this desire of its constituents and does as little as possible to limit personal freedoms. While I was in New Mexico, I sat down with Assistant Attorney General Frank Weissbarth. He could not have explained the situation in more neoliberal terms. His specialty was consumer protection and subdivision law, so he was familiar with the issue of colonias. After I left New Mexico, he would win the public lawyer of the year award from the state bar in New Mexico in 2006. He believed less regulation was "the culture of New Mexico." The real issue, as Weissbarth saw it, was housing and how to "provide housing for families that make $15 thousand or less [a year]." According to Weissbarth, "They don't fit into our mind set of who lives in America."[13] When I asked him if he thought things were any different for white working-poor families in town compared to the Mexican farm worker families in the colonias, he did not think so. Race was not a key issue in Weissbarth's mind, and he did not appear comfortable discussing the topic. He did mention the case in Hatch when Mexican residents of the trailer park were forced into the colonia (Recuerdos) and explained the justice department of New Mexico did call this racial discrimination. But in this case, he said it was in terms of where trailers can go based on property values. In general, though, race is less an issue than property, and developers are simply looking to profit off the poor. While this desire for freedom and privacy is primarily associated with the western United States and the American desire to escape an interventionist state, it also fits with the history of northern Mexico where the border region's distance from Mexico City made it a refuge for rebels and revolutionaries of all sorts.

In relation to the El Paso area, Larson describes the colonia situation as follows: "The combination of limited governmental power and a political culture of resistance to regulation of private property meant that until only recently, El Paso County had neither the legal tools nor the political will to respond to such substandard conditions" (1995, 198). This described the situation in New Mexico in the 1980s and early 1990s as well. Interestingly, Larson is clearly describing the political project that is neoliberalism without using the term itself: "Limited governmental power and a political culture of resistance to regulation of private property" perfectly describes the shrinking of the state and the valorization of privatization, two central aspects of neoliberal policy. Unfortunately, this combination meant counties on the border had few resources to go after colonia developers who created illegal subdivisions to sell to immigrants. Weissbarth saw colonia development in New Mexico as "symptomatic of America . . . We feed on our poor people now. Colonias are symptomatic of that, the flip side is people can't afford a house."[14] He blamed the situation on Ronald Reagan, who he believes convinced the country that it was a bad idea to subsidize housing. According to Weissbarth, the state indirectly ends up spending millions to fix the bad housing that private industry creates in the form of colonias. In the long run, this second-tier cost is higher than the cost to create subsidized housing.

When our conversation was nearing an end, Weissbarth summarized his version of the Reagan-era housing policy: "It is not the state's place to provide housing so housing subsidies were cut and you had to find another way to buy a house. Because that's what real Americans do, they buy a home, no matter how much financial hardship it brings them."[15]

In New Mexico, along with the stress on owning a home, the lack of a solid subdivision law led to the easy development of colonias. Colonias developed throughout the 1980s and early 1990s with little or no response from either the county or federal governments. The Doña Ana County government knew there were colonias, but, with its limited budget and human resources, it could not keep track of them. Doña Ana County is one of the poorest counties in the country, and, with its low tax base, it has few, if any, resources to contribute to addressing the "colonia problem." It was not until 1989, when the federal government signed the Cranston–Gonzales National Affordable Housing Act, that Doña Ana County realized it would have to do more about colonias. The Cranston–Gonzales Act defined the term "colonia" and identified New Mexico as a state in which colonias existed (Paster 1993, 1). Yet it was not until 1992 that the county formed a colonia task force to handle the colonia problem. It quickly became clear that the lax subdivision laws in New Mexico were to blame for the ease of development of colonia communities. The task force focused on how to improve the subdivision laws and how to distribute the growing federal monies for colonia improvement.

Until 1995 when the laws changed (after years of battle in state courts),[16] creating a subdivision in New Mexico with little or no infrastructure was

relatively easy. The current law defines a subdivision as a "division of a surface of land, including land within a previously approved subdivision, into
two or more parcels for the purpose of sale, lease, or other conveyance; or
for building or private road development, whether immediate or future."[17]
Before the new law was created, State Statue 47-6 defined a subdivision as
being the division of land into five or more parcels within three years for
the purpose of sale or lease. Based on this law, it was entirely legal to split a
parcel into four lots, sell these, and then split those four into another four
after three years. Many developers did not wait the regulated three years
but went ahead and split the land a second and third time immediately. An
intermediary often enacted the second- and third-level splits so it was not
obvious that one person still owned the land. The law itself set no clear
penalties for violations, making it difficult for the county to prosecute the
developers.[18] Developers often bought arid farmland from farmers at next
to nothing and resold it to colonia residents for a great profit.

Doña Ana County lawyers experience ongoing problems bringing developers to justice because of the method through which developers sell colonia lots. Most colonia lots are sold under contract-for-deed agreements.
Contract-for-deed is a legal way to sell land in both Texas and New Mexico
and is very popular because it does not involve the government in any way,
thus insuring privacy and reaffirming colonias as self-sufficient communities
(Ward 1999). Under contract-for-deed agreements, the buyer purchases land
from the seller, who is also the financier for the sale. An agreement is drawn
up that usually involves long-term mortgage payments at very high interest
rates (up to 35 percent). These payments go directly to the seller, who has
the power to repossess if even a single payment is missed. The deed is only
given to the buyer upon completion of the contract when the mortgage is
fully paid. It is only at this time that the county records the sale. Because of
the use of contract-for-deed agreements in most colonias, the county has no
certain knowledge that the agreements even exist until the first buyer has
paid in full. By this time a colonia is already established, and there is little the
county can do because the buyers are already long-term residents.

The county often becomes aware of colonias through the daily workings of county offices. Tax collectors and assessors may not know where
colonias are, but the county geographic information systems (GIS) lab
can usually give you a pretty good idea of where to look. When I went to
get base maps for my research, I found the GIS lab already had fairly accu-
rate maps of the colonias in the county. The head of the GIS lab explained

to me that these maps were necessary to provide emergency medical services and other emergency services, such as the police and fire depart-
ments, guidance on how to get to some of these communities when called.
Colonias exist in rural areas where there is no postal service, so mail delivery does not equate to an address. The first time most colonias residents
must face the fact that they do not have official physical addresses is when
they require a delivery or in an emergency. According to the GIS lab, 911
calls "outed" several colonias over the years. Although colonia residents are
unlikely to call 911, their communities do receive street signs soon after discovery in order to aid future emergency responders. Flora explained the fear
of dialing 911 that many colonia residents have. She said, for example, if you
hear your neighbor beating his wife and you want to stop him, you cannot
call the police because that might lead to his arrest *and* that of his wife if *La
Migra* (border patrol) follow the police and find that she is undocumented.
Since it is impossible to know the documentation status of all one's neighbors, one simply cannot run the risk of calling law enforcement officials or
any emergency services into the community because *La Migra* tends to follow. It is illegal for the border patrol to follow the police, which is commonly
known as "piggy backing," but this occurred when I was in the colonias
often, even though the laws that forbid the tailing of legitimate police calls
exist so people will not be afraid to call the police in situations just like this.

It was not until 1995, after ten years of battles led by colonia activists,
NGOs, politicians, and lawyers, that legislators rewrote the New Mexico
subdivision laws to make new colonia development illegal. Yet because
this new law applies only to border counties—a stipulation that had to be
included in order to get the law passed—colonia development now occurs
in the interior of the state. The battle to change the subdivision law was
long and drawn out. Many residents of New Mexico felt they would lose
the freedom to subdivide their land. This was particularly true of ranchers who wanted to be able to split their land between their children, and,
in the end, an exception was included to allow family-based splits. Frank
Weissbarth described these exceptions as "potentially dangerous loop-
holes" through which more colonias could be created by family transfer
of property. He was also concerned that the 1995 law had no real oversight
process. As long as affordable housing is scarce on the border, money will
be made by the sale of colonia lots, and evasive practices will continue.

Although the state of New Mexico tried to stop the development of
colonias, its inability to do so, accentuated by the strong desires of many

residents of New Mexico to limit the role of the state in the subdivision process, has sanctioned continued development of colonias. The ambiguities of the role of the state in colonia development deserve a closer examination.

The original impetus to revise the subdivision law closely related to the growing realization that colonias posed a large-scale problem and that the lax subdivision laws were to blame. At this time, during the early 1990s, a group of colonia activists, democratic politicians at the state and county level, and lawyers in the New Mexico attorney general's office began to lobby the state legislature to change the laws. This coalition aimed to create a very firm subdivision law that would make the future production of colonias impossible, or at least easily prosecutable. For many of those active in this political battle, the colonias were, and are, inadequate sources of housing and discriminatory because they targeted the population of Mexican working poor who were desperate to own land and unfamiliar with their rights as U.S. residents and citizens. This is a population that often finds access to rights and representation difficult due to language barriers and inadequate knowledge.

While some groups worked against the production of more colonias, other forces at the state, county, and community levels fought back, defending their ability to both subdivide and sell their land to whoever was willing to buy it. It is in these contradictory forces at the state level that the influence of neoliberal politics becomes clear. The arguments of both the developers and state officials against the 1995 subdivision law closely relate to neoliberal rhetoric of self-reliance and self-determination, culminating in arguments around basic rights to private property and the ability to make a living (Nelson 1995). In this way, the production of colonias in the very first stages of subdivision can be viewed as rooted in neoliberal ideology.

Lack of Affordable Housing, Border Style

No one buys an unequipped colonia lot on purpose; that is to say, all colonia residents buy their land believing the developer's promises that infrastructure is on the way. Women leaders of the colonias described a variety of reasons why they purchased their lots, but a few themes did stand out. The first was price. They were all looking for land that they could afford on a farm worker's salary. This is a common reason given to explain the emergence and continued growth of the colonias, so I expected to hear residents discuss the low price of colonia lots as an incentive to buy them. The second theme was less expected, but just as commonly mentioned: it was a sort of "cultural and economic imperative" to own land. As Mexicans, the women explained, it was very important for them to be able to buy land and hand it down to their children. This was the "Mexican Dream." As Flora described, "I wanted to have something of my own. It was not pretty, but well, it was ours. I didn't want my kids to always be dependent on someone else's home . . . Here I have lived the happiest days."[19] Here the "Mexican Dream" looks a lot like the "American Dream," and both mesh with neoliberal logics of self-reliance and independence that dictate the individual good and goals are more important than the social good.

Colonias offer an acceptable alternative to renting for several reasons. Colonia lots are cheaper monthly, and colonia residents need not pass a credit check to rent property. For Esperanza, her colonia lot proved to be the cheapest option: "It [a colonia lot] was the cheapest. We did not have the money to get expensive land. We liked a lot in the forest, there in the forest they have nice lots; but the monthly payment was very high and we couldn't afford it. My husband almost never lasted in his jobs. He left work because of his illness . . . and because of all that we got this land that had cheaper payments."[20] Peter Ward explains the lack of public housing on the border through an examination of the prevailing American attitude about housing. Federal and local governments focus their attention on creating incentives to home ownership rather than creating public housing for the working poor. Ward comments that "indeed public housing is stigmatized in the United States" (1999, 87). The stigma Ward describes exemplifies the stress on personal responsibility, one of the basic elements of the ideology that sits at the foundation of the neoliberal political project. The true neoliberal citizen should be able to provide for her own housing regardless of her impoverished position.

Larson argues that the growth in the region's employment opportunities and lack of affordable housing led directly to the use of colonias to house the area's labor: "In the 1970s and 1980s the dearth of affordable urban housing, combined with the intensifying appetite of the local labor market for workers, created a demand for low-cost housing alternatives in the region. Despite public expressions of concern about health and housing conditions in the colonias, it seems clear that the border economy relied upon these settlements to keep alive a steadily more impoverished workforce for two decades" (1995, 220). The two decades Larson describes have become more than three decades, and there is no end in sight for the use of colonias as the primary source of affordable housing in the area.

The colonias are a textbook example of the market leaping into the breach and providing resources the state chooses not to provide. Between the cunning of developers out to make a profit and the resourcefulness of the Mexican working poor, colonias were created as an alternative form of housing along the border. With no credit history in the United States and little collateral, Mexican immigrants often cannot access the traditional homebuyer's market. After the down payment, monthly payments for colonia lots are often less than local rents. Most of the families with whom I worked paid between $150 and $300 monthly. But once the monthly payments are made, the working poor who inhabit the colonias have very little money left to pay for the necessary improvements to their lots. Self-help projects can help to solve this dilemma. In this way colonia residents build their own homes, install their own waste-water systems, and improve their community's roads with their own labor. In self-help projects, residents work alongside professionals to learn techniques, to defray cost, or simply to build it themselves with loaned equipment. Working as individuals, families, or neighbors, colonia residents have done enormous amounts of work to improve the conditions in which they live. My findings support those of Larson, who argues that this employment of self-help projects and technology is a logical outcome of the conditions in which the working poor find themselves (Dolhinow 2001):"Squeezed between falling wages and declining public support for housing programs, it can be predicted that working poor households in the United States increasingly will turn to self-help housing to survive" (Larson 1995, 246). Self-help is a common method of resource provision in Mexican colonias, and because of this it has been argued that it is familiar and therefore easier to live in U.S. colonias for Mexican immigrants. But a cultural familiarity with self-help alone is not enough to explain the development of colonias.

The Production of Colonias

> Cecil McDonald's Texas Colonia Creation Recipe
> Works for New Mexico as well, just add two portions more
> dust.
> Makes 20 or more slices . . .

1. Preparation: Perform market research. Find a poor community with conservative banks that will not provide mortgage financing for migrants or recent immigrants and which also has a shortage of housing for low-income families.
2. Find a willing attorney to research land development, septic tank, and water supply regulations. In the absence of those regulations, you have a good potential for developing a colonia.
3. Select an area close enough to but at the same time away from the city where building and development activities are not noticeable.
4. Negotiate with the landowner, offering to pay double the asking price as long as the sale is owner-financed. Offer to pay 10 percent down and the rest in a very short term at a negotiable interest rate; the owner has already made money doubling the price of the land. Do not get any banks involved. The transactions are to be kept secret; you don't want to lose the element of surprise. Fiercely protect the anonymity of the landowner. Sell lots on a contract-for-deed basis.
5. Get a surveyor or engineer to develop the subdivision on paper.
6. Advertise, but not in the nearby city. Ask for a very low down payment, even as low as $25 or as little money as the potential purchaser has in his pocket. Starting next month, he will have to produce $125 or $150 until he pays off his debt.
7. By the time local officials realize a colonia is developing, hundreds of lots may be sold and many families may already be living on the site.
8. By the time there is an official local action or state legislation enacted, the colonia will likely be completely sold.
9. If pressured by the state or county to provide improvements such as water, sewers, or streets, argue lack of funds but continue selling lots in order to finance the improvements. Meanwhile, keep on making money. It may take a year to be subjected to enforcement action, at which time you declare bankruptcy because of the high costs of improvement.
10. Relocate to another city. Communicate with attorneys by mail.

The most important ingredients in this recipe are time, distance, and surprise. (Ward 1999)

Cecil McDonald is a well-known colonia developer in Texas. If any-one knows how a colonia is created, McDonald would; his knowledge comes from experience. McDonald created this "recipe" in Texas, but this recipe works in New Mexico as well. In New Mexico it might be "Ysidro Lopez's Recipe for Colonia Creation" or "Ken Anderson's Rec-ipe for Colonia Creation." The names might differ, but the methods and motives are the same. As McDonald's recipe makes clear, a very simple method defines colonia creation, which takes advantage of some of the most obvious conditions along the border: poverty, a lack of affordable housing, and minimal protection for recent immigrants. The lax sub-division laws that used to be in effect in both Texas and New Mexico made colonia creation that much easier. In addition, because most bor-der counties are large and relatively uninhabited, county officials rarely have enough employees to monitor the development that is taking place in and around the edges of the county's jurisdiction. Finally, the more actors involved in the creation of a colonia, the better; that way, the pri-mary developer can disguise his participation. As McDonald and other developers have made clear over the years, anonymity is the key to colo-nia development. Perhaps the most important ingredient in McDonald's recipe, though, is the tone. It appears to be a joke—a very offensive joke, but a joke nonetheless. Yet, McDonald is serious. This is how McDon-ald and others make money at the expense of immigrant populations. McDonald himself stresses the importance of the secrecy that makes developers money and makes it so hard to catch them.

When I described how developers create colonias, covertly splitting lots and spreading the responsibility for the misdeeds over many dif-ferent culprits, some who had no idea their name was on a subdivision agreement, I often use the example of a famous 1970s shampoo com-mercial. Some readers may remember the very catchy ads for Faberge Organic Shampoo staring Heather Locklear. I was quite young when they were in heavy rotation, but they had a lasting impression. The ad featured a fabulous-looking Locklear tossing around her extra-full, blonde locks, exclaiming that if she told two friends about Faberge Organics Shampoo and you the viewer did the same and then they did the same and so on everyone would soon know and we would all have great hair. All the while the screen filled up with pictures of gorgeous blondes as if she was getting the word out about the shampoo while we the viewers were watching. This is exactly how you make a colonia, with the only differences being that you have to keep it a secret and most of the time those involved are not nearly as glamorous.

A lack of accountability is perhaps the most important characteristic of this recipe. Through fragmented networks and a decentralized organization, colonia developers keep their own roles hidden yet reap the benefits of their investments. As if knowingly deploying Foucauldian definitions of power, colonia developers create "the net-like organization" of which Foucault speaks when he says, "Power must be analyzed as something which circulates, or rather something which only functions in the form of a chain" (1990, 98). In this way, power creates a truly fragmented and circulating power structure, one that is also very difficult for county attorneys to prosecute.

More often than not in New Mexico, the land purchased by developers to subdivide into colonia lots is arid farmland. Once developers subdivide the land, they usually sell lots by word of mouth or via signs posted on the rural roads that serve as access to the colonias. Migrant farm workers drive these back roads to and from their work in the fields every day. Many of the colonia families I met found their future homes while driving around the back roads in search of "land for sale" signs.

Few people know the ins and outs of colonia creation as well as the Doña Ana County lawyers who try to prosecute the developers. Through years of working on colonia cases, these lawyers discovered the many secret and illegal mechanisms developers use to obtain land, subdivide it, and sell it to unsuspecting Mexican immigrants. Karen Acosta served as the lead colonia lawyer in Doña Ana County while I was there, and she and I discussed at length the issues on which the county focuses when it tries to prosecute a colonia developer: "The number-one, fact-based problem [she encountered working on colonia cases] was figuring out who owned it [the colonia], when it happened, and now who thinks they own the land. We know who was the owner of record at one time, but it may have passed through a lot of hands after that."[21] One of many traps developers set for lawyers is the creation of fake owners for subdivided lots, a process that makes tracing colonia ownership next to impossible. According to Karen Acosta, Ysidro Lopez, one of the main developers in Doña Ana County, was an expert at hiding his part in the colonias he developed:

I think there are a lot of contracts that he has on the land under other people's names where he really owns it. A woman called just because she saw her name in a publication notice in the paper of the assessor's office to sell some land, and she says, "I don't own that anymore. I bought that from Ysidro Lopez years ago, and there were so many problems with this, that they didn't have access and all those kind of things, that I just walked away from it. I just went to the bank, or whoever was doing the financing at that point, and said, 'I don't want this any more,' and they said, 'Fine.'" Now it's still in her name. Well, he doesn't have an incentive to get it back in his name officially by our records because it's going to look like he owns it.[22]

Considering all the other demands on their time, colonias being only one of many projects the county lawyers handle, the secrecy and subterfuge that surround the development of colonias make it difficult for the county to make progress in forcing developers to pay for the necessary improvements.

Colonias Here and There

Colonias, unplanned subdivisions, exist on both sides of the international boundary.[23] Those in Mexico have a longer history and are more common. Mexican colonias function very differently. As several studies demonstrate, colonias are accepted alternatives to overpriced housing markets in Mexico (Staudt 1998; Ward 1999). In the ever-expanding colonias of Juarez, Mexico, squatting and informal takeover are the most common processes through which these communities are initiated. Once the land is "settled," the Mexican government usually takes notice and begins to supply basic services. Local governments improve conditions in the newly settled colonias because improvement might garner votes for the next election, and electoral politics in Mexico are often about patronage (Rubin 1997, 11). Groups of citizens who believe that, with time and a little political work, their colonia will be given services by the state intentionally create colonias in Mexico. The question has been raised in relation to U.S. colonias: did colonia residents believe that the government would supply them with basic services over time because this how it works in Mexico (Paster 1993, 6)? If this is the case, then the choice by colonia residents in the United States. to buy in a colonia takes on different meanings.

In the case of my sample, the women expected that the developers would provide the necessary infrastructure, as promised, soon after the colonias were opened to residents. Colonia residents were justifiably upset by the broken promises of developers, but they often live with this anger in a resigned way. Women who become leaders tend to channel this anger into their activism. But even the leaders sometimes join their fellow residents in a quiet sort of resignation. As Mexican immigrants, colonia residents are not strangers to discrimination. Rather than dwell on their misfortunes, colonia residents focus on the future and on the improvements they have made to their communities and are planning to make. This is not to say that colonia leaders are not excited by the county taking the initiative and attempting to sue developers and legally force them to provide infrastructure. Colonia leaders love to see developers brought to justice. But experience has taught these women that the developers are usually too cunning and the county too understaffed to ever really create change in the colonias through legal channels. Fair or not, improvements usually fall to colonia residents.

The self-help ethic, discussed earlier in this chapter, is both common in Mexican communities and a crucial aspect of neoliberal interventions in the colonias. They often build their own homes slowly, over time, room by room. Unlike Anglo-American communities that frown upon slow building techniques, the sight of a house under construction for decades in a colonia does not merit calls to the police, the permits department, or both. Coming from a background where self-help is a common and valued way to get ahead, colonia residents are comfortable using these methods to improve their communities in the United States.

The lack of state-sponsored, affordable housing combined with lax subdivision laws led to the creation of colonia communities. A strong neoliberal presence and policy agenda underlies this chain of events. Self-help rhetoric and techniques not only provide colonias with basic infrastructure but also do so in ways that make these absolutely necessary improvements logically appear to be the responsibility of colonia residents. The techniques of governmentality employed by most NGOs in the colonias are so deeply embedded in neoliberal ideology of personal responsibility that self-help has become the primary method currently employed. These connections between NGOs and neoliberal discourses of self-reliance and individual attainment are maintained via funding initiatives, donor visits, follow-up reports on grants, quarterly reports, and many other forms of disciplining that NGOs experience at the hands of their funders.

Culturally Familiar Living and the Production of Colonia Residents

Colonias provide Mexican immigrants with a "culturally familiar form of living," which allows people to anchor themselves culturally in what might otherwise be a very foreign setting. But this culturally familiar living serves another purpose as well: it serves the goals of neoliberalism. Duggan points to the "rhetorical separation of the economic from the political and cultural arenas" in neoliberalism's daily working to redistribute wealth upward (2003, xiv). Neoliberalism uses cultural identity, sexuality, race, and religion to implement its politics. In Duggan's research, she found that the gay rights movement and homonormativity were appropriated by the neoliberal political project to "actively obscure" its own work within that community (3). Based on the cultural familiarity the colonias offer through their likeness to rural living in Mexico and most residents' history with self-help techniques, it appears a form of consent is created for life without basic services in the colonias.

According to Sylvia, a former colonia resident and activist, "Colonias are almost completely Mexican communities because this is the only way these people have a feeling of home. If you ever visited any of the communities they came from in Mexico, you would see the sameness and the feeling of family and friendship that you feel in colonias today."[24] Sylvia lived in a colonia with her husband for years before they saved enough money to buy land in town. She was an activist in her colonia, and, through this experience, she gained employment working with several NGOs as a translator. Although no longer a colonia resident when I met her, Sylvia was still involved in the colonias as an employee of several NGOs.

The vast majority of colonias house exclusively Mexican immigrants and it is these immigrant residents who describe their communities as "Mexican." As Sylvia said, these are "Mexican communities," and I met very few residents who would identify as Mexican American, including second-generation residents who were born in the United States. The majority of households are recent immigrants, primarily first and second generation, and they choose to live in the colonias because they are Mexican spaces. Thus, they freely and openly identify as Mexican in these spaces. Outside the colonias, "Mexican" may be a term with many negative connotations (Vila 2000), but, within the colonias, being a Mexican is a valuable and even necessary aspect of one's identity. It was a running joke that if I were not a researcher I would not have been able to stay in the colonias because I am not Mexican.[25]

Yet, colonia-like communities exist across the country wherever the working poor seek alternative forms of affordable housing. Other groups of working poor share some of the characteristics, such as the need to purchase lots while having low or nonexistent credit ratings, that make colonias a good housing choice for Mexican immigrants. Other characteristics are more specific to the Mexican immigrant population, such as their background with self-help projects. My project situates the colonias in New Mexico within the particular historical and geographic juncture in which they developed and stresses the cultural aspects of colonia living.[26] Immigration to a new place can be a very unsettling experience. Even years after their arrival, the immigrant women with whom I worked still seek a community that provides familiarity and a connection to Mexico. Sylvia describes this familiarity in practical terms: "In Mexico you can visit your neighbors without having to get in your car, you can share resources and borrow a cup of milk from anyone in the community," she says. "You don't need a phone because all you have to do is stick your head out the door and shout for your neighbor and she will hear you. It's like being in Mexico all over again and that connection is necessary for the people that come from there."[27] Colonias provide a cultural anchor when the rest of everyday life differs dramatically. Colonias keep intact Mexican cultural meanings, and recent immigrants can find this familiarity comforting.

Buying land in a colonia community also offers benefits other than the accepted do-it-yourself construction. Colonia lots are often large enough to house more than one household or can be bought side by side so that related households can live in proximity. Extended family is very important in Mexican communities, and the preferred housing arrangements are those that allow extended families to live together. Most of the women in my sample had extended family living in their colonia, on either the same lot or those nearby.

In Estella's case, shortly after she arrived in Recuerdos, her father and mother moved onto the lot directly next to hers. Now retired after many years as an agricultural field worker dating back to the Bracero program period, Sr. Perez spends his time driving his wife and grandkids around town and adding on to his home. Estella's parents run a very popular and successful store out of their back door. A few years back, Sr. Perez built a small laundry room off the back of their trailer. He later added a split door, and now he and his wife sell candy, sodas, and chips out of their laundry room through the top half of the door. While I was working in Recuerdos, Sr. Perez enclosed the long porch across the front of the trailer to use as a television room and semi-outdoor eating area.

During a follow-up trip to New Mexico in October 2001, I visited Flora and her mother, two of the most dynamic colonia leaders in my study. Flora was temporarily living on the edge of Troy, a town just south of her colonia. Flora was residing there because her long-awaited new house was under construction on her lot in Los Montes. Flora was very excited to be one of the first colonia residents to get a new house funded by the federal government and a local nonprofit housing corporation. For very reasonable payments, under three hundred dollars a month, Flora would get a three-bedroom house. The plan on the day I came to visit was to meet in Troy and then drive to Los Montes and watch the foundation being poured for the new house. When I arrived in Troy, I found that most of her extended family had come down from Los Montes to greet me. Flora, two of her sisters, her mother, and various grandchildren were all packed into the kitchen making coffee. Once we had the obligatory coffee and *pan dulce* we all headed up to Montes, where we had more coffee and pan dulce, this time in the trailer of one of Flora's sisters. While we were sitting around talking in came another sister with her children. At this point, the four sisters, the mother, and the many grandchildren, who all live in Los Montes, decided to call the one sister left in Mexico and wish her a happy birthday. With one sister sharing her lot, another living with her in-laws down the street, and a third sister around the corner, Flora has an extraordinary support network in her community.

The space for extended family is only one of the ways in which colonias attract Mexican immigrants. Most colonias residents also expressed a desire to live with other Mexicans. The women in my study often described their Mexican communities as comforting and familiar. "I say [I like Mexican communities because] in order to feel in my country. Because there is nothing like living with other Mexicans," Marie said.[28] Esperanza likewise told me, "We get along well with each other. It's more than we all speak Spanish, we all know each other. We have a lot of Mexicans here, and we get along well. I am happy here."[29]

The women described their colonias as places where their children can grow up Mexican in the United States. Several women said they believed their children were safer in a colonia with other Mexican families. In other mixed communities, said Esperanza, "We don't get along as well. And problems begin with the kids, this and that happen. And here with the Mexicans this and that do not happen."[30] Distance from the cities also benefits children because it keeps them out of the kinds of trouble more urban children experience with gangs and drugs. Politically, a predominantly Mexican community also offers benefits to recent immigrants. The harassment Mexican immigrants experience at the hands of the border patrol can really only be understood by others who have experienced similar treatment. In colonias, a wariness of the border patrol and an understanding of the need for privacy in matters relating to the U.S. Immigration and Customs Enforcement is common sense.

The women I interviewed found the isolation of colonias a benefit that provided the privacy and peace they desired. Many of the women came from a generation that grew up on rural ranchos in Mexico. They want their children to have a similar experience. Past experiences in apartment complexes in the United States were unpleasant for them and only increased their desire to have their own land. For them, colonias offer the rural feel they grew up with and a way to make their children's youth a bit more like their own. Estella often spoke to me about her desire to live in a rural space like a colonia and not in the city. One of her sons and a daughter were both living in the nearest big city, and she visited them fairly often, but she much preferred when they came to see her. She would tease me about my life in the big city and tell me that after a year in the colonias I would never be able to go back. In describing her own life in Recuerdos, Estella said, "We began to live contented and to enjoy how it smells in the mornings here, that it smells fresh, that there are women who have chickens and roosters that sing and you hear them, and that you can get up and walk all around without having the neighbors all over you."[31] For Estella, her colonia offers all the nature and peace of her Mexican childhood and a degree of separation from her neighbors that traditional subdivisions do not.

As Estella mentions, the space that makes colonias such good places to raise children also offers a relatively high level of privacy. Alicia is a high school Spanish teacher who values her privacy: "I mean I'm sociable, but my home is my castle . . . I don't know how to explain it, but to me, you know, whatever is outside really doesn't matter. I like living away from the houses. You know, I don't like one little house and then you turn around and your neighbor is right next to you . . . I don't like that closeness. I like my space; I need my space."[32] It would be wrong to think Alicia is a recluse;

she is very active in her colonia and knows many of her neighbors. She just does not want them to know everything about her life. Colonias are communities full of friendly and helpful neighbors, but they are also communities with a great respect for privacy. Colonia neighbors do not meddle in each other's affairs. They gossip endlessly, but they rarely interfere. If a resident fails to mow her lawn for a month, neighbors will not reprimand her, though they might come over and help, or at least lend moral support, when the mower is finally brought out.

Larson points to the ownership of land in a colonia as a "powerful symbol of self-reliance, personal dignity, and family advancement" (1995, 207). She also comments that "the names given to El Paso County subdivisions and to the streets within them express these aspirations. Colonias named Las Azaleas, Las Dahlias, Bosque Bonito (Pretty Forest), and Joya del Valle (Jewel of the Valley) evoke lush, pastoral beauty in these parched desert settlements" (207). The colonias in which I worked have similarly peaceful and nature-inspired names. "*Tranquila*" is a word most all the women used to describe their lives in the colonias. *Tranquila* translates to more than just "tranquil"; it implies a sense of peace and an easygoing pace, a calmness that many of the women I met desired in their lives. For example, in a conversation about her favorite aspects of living in Recuerdos, Juana said, "What do I like best? I like the peacefulness [*tranquilidad*], that there is not so much moving around, not so much running around—zoom, zoom, all the time. I like it for my kids that there is so much free space, there aren't fights, there's nothing like that. It's a peaceful thing. That's what I like about it."[33] Colonia residents so value this quality that they are willing to live without basic services or will provide them themselves when the state abandons them.

The Production and Construction of Colonias Spaces

If we take the body as the fundamental scale at which space is experienced, it becomes clear that everyday situated practice executed using the body contributes to the social construction of space. Yet before the activism of colonia leaders reproduces the spaces of colonias, these spaces must be produced for Mexicans to inhabit in the first place. In proving his claim that "(social) space is a (social) product," Lefebvre argues that "every society—and hence every mode of production with its subvariants (i.e., all those societies which exemplify the general concept)—produces a space, its own space" (1996, 31). As I have shown, the colonias emerged in order to house the growing numbers of Mexican working poor that were and are attracted to New Mexico's agricultural economy. For agriculture, as a capitalist enterprise, to succeed, it must have plenty of cheap labor at its disposal. Colonias provide the least expensive way to house workers near the capital that requires them. According to Lefebvre, "To change life, however, we must first change space" (190). Here, Lefebvre's arguments begin to lead to a discussion of the importance of social reproduction in the processes of globalization. For colonia leaders to create change in their communities, they must first change the spaces of their communities, according to Lefebvre, and, for the women in the colonias, they do this through their work in the processes of social reproduction.

Mexican Spaces

Colonias police the boundaries of "Mexican-ness" in southern New Mexico. Not only do colonias keep poor Mexican immigrant labor out of sight, but they also keep the most recent Mexican immigrants out of more Anglo and mixed neighborhoods (Nevins 2002). Although money primarily motivates colonia developers, the communities they create are also fueled by racial discrimination.

Donald Mitchell discusses "race" as a strategy for the organization of access to rights: "It is in this sense that 'race' may be understood as a 'strategy' for advanced capitalist societies: if race can be seen as a natural basis for unequal access to rights . . . if race can be presented as a commonsense reason why some in society are meant to be more privileged (with money in the eyes of the state), then such inequality is made to appear pre-ordained, as simply the natural order of things" (2000, 256). On the border we see discourses created that justify Mexicans living without basic resources because they "like" to live that way, as if poverty and filth were somehow the "natural" state for Mexican immigrants. But the relationship between "Mexican-ness" and rights in the United States embraces more than just race and ethnicity. Citizenship, and all it entails, is also centrally implicated in the discussions of Mexican immigrants and their living situations. Through these discussions of citizenship and rights, the state and employers can justify the inequality they propagate. When local discourses dictate that Mexicans like to live in poverty and that they feel more comfortable without basic services, it becomes much easier for neighbors, politicians, and employers to accept the conditions in colonias and rest easy knowing that they do not need to be concerned because this "inequality is . . . pre-ordained" (256).

Along with the efforts of capitalists and developers and the other processes of the social production of space, exist the processes of the social construction of space. In the case of the colonias, these spaces of resource deprivation are constructed from discourses both inside and outside the communities. Because these communities are so closely identified with their Mexican inhabitants, many of the discourses surrounding them relate to their ethnic identification:

> In El Paso, as in other areas of the country, the discourse of race and ethnicity is pervasive. Nevertheless, here it combines with a discourse of nationality in a volatile mixture that, for many people, marks almost anything that is stigmatized as Mexican. Poverty is named in Spanish in El Paso—and in El Paso, Spanish signifies Mexican. For instance, the poor neighborhoods of the city are not known as neighborhoods, slums, ghettos, or shantytowns but as "colonias." This is the case irrespective of the language spoken or the ethnicity of the speaker . . . In this context, the use of Spanish kills two birds with one stone, for Spanish is both the language spoken in Mexico and the cultural marker for a great part of the Mexican American community in El Paso. Consequently, the use of Spanish to name poverty in El Paso is part of a local hegemonic discourse invoking both national and ethnic classification systems. (Vila 2000, 83)

The name "colonia" carries the same connotations in New Mexico as in Texas. Outsiders similarly construct it as "Mexican" and therefore "impoverished." Poverty is associated with Mexican immigrants not because they are immigrants, because they are field labor, or because they are poor and uneducated, but because they are Mexican. Along the border the national identity of "Mexican" is constructed in opposition to that of "American." Vila demonstrates how identity, ethnicity, poverty, and place all mesh together in the social construction of colonias as racialized spaces.

Just as the Mexican immigrants who live in colonias construct their communities, they also construct their own identities as colonia residents. They construct these identities in opposition to those constructed from outside the colonias. The women with whom I worked view their communities as "Mexican" but not impoverished and undeveloped. Rather, they take pride in their colonias and concentrate on the improvements they have made. When asked if they would ever consider leaving their colonias, all the women said no because they have made these poorly planned communities their homes through their own labor. In Montes, Flora and others have planted trees and gardens that they cultivate with the well water they cannot drink or with the grey water from their washing machines. With a little green and shade, these women in the middle of the desert transform a patch of dirt into a garden. As more and more greenery shades Montes and Recuerdos, the colonias look less accidental and more realized.

Conclusion

The ambiguities of colonia communities exist on many levels. On a very basic level, colonia developers take advantage of recent immigrants who do not know the legal system and have few resources. But at the same time, without colonia lots this group of working poor would not have any land to buy. A similar ambiguity characterizes the activism of colonia leaders. If the women who lead the colonias do nothing, they and their families must live without basic services. But if they do organize and inspire the colonia to build its own waste-water system or grade the roads, they appear to give their consent for the state's abandonment of their communities and the conditions in which developers left them as well.

My central argument revolves around the set of forces that make the production of colonias possible. The interrelated processes of global economic restructuring, international immigration, and neoliberal governmentality work together in the production of the colonias. When placed geographically on the U.S.–Mexico border, these processes intersect with the long history of movement between the two countries. It is at this site of convergence that the conditions are created, and, thus, the production of colonias appears both necessary and useful. At this particular historic and geographic conjuncture, a triad of key actors has developed to make production of colonias as neoliberal spaces feasible. These actors include colonia developers, the state (i.e., local and federal governments), and NGOs.

The production of colonias as neoliberal spaces is important because it is from this foundation that colonia activists and leaders arise. The daily presence of the neoliberal political project and its attending ideology and discourses, as presented by the state and NGOs, has a strong effect on which women become leaders and how they enact their leadership.

CHAPTER 2

The Production of Women
as Neoliberal Leaders

Mejor solo que mal acompañado.
Better alone than poorly accompanied.

"*DILE PINCHE CHILE.*" I knew I was almost as red as the chile I was desperately trying to manipulate into a ristra when Estella made this off-color comment. At this point I had been working in the colonias for almost nine months, and I had a pretty close relationship with Estella. (Or should I say she had one with me?) Estella is a commanding woman and can be a very intimidating presence. When it came to ristra making, she was the queen.

The group of us—Estella, her mother, myself, two or three of her cousins, and at least five nieces and nephews, depending on the day— were meeting in mid-September to make ristras as I described in this book's introduction. By the time the chile harvest is in full swing, front porches across the Hatch valley become small-scale ristra production lines. In the colonia of Recuerdos, none is as lively as the one run by Estella and set up on the front porch of her parent's trailer on the lot next door to hers (see Figure 2.1).

So how did Estella end up speaking so poorly of my chile on my first day making ristras? As with many things involving Estella, there were a lot of jokes told while making the ristras, and more than a few were colorful. She had a large repertoire of sexual jokes, and she loved to tell them. This made her mother quite uncomfortable, and I saw her blush intensely as the day went on but could also hear her laugh in spite of herself. Estella's father, on the other hand, who helped with the crates of chile, tended to stay at the edges of these conversations. These were women's spaces, both in work and in jest.

All the while, I labored diligently at my ristras, trying to find the magic tension between too loose where the chile falls out and too tight where it breaks right off, and trying to follow the conversation and jokes going on

Figure 2.1 All in a day's work: Ristras lined up on Estella's production line

around me. It is true my Spanish had improved vastly in nine months, yet my years of Spanish classes could never have prepared me for the jokes, puns, and sexual innuendoes Estella had waiting for me. What I could do was talk with the whiny two-year-old next to me. She was upset because her mother was talking to Estella and not to her. So I made a joke with her and told her to tell her mother what her mother kept telling her: "Mommy, hush, hush." Jokingly, I encouraged her to tell her mother to be quiet. As children that age do, she parroted the phrase over and over and we all laughed. Not long after, I pulled too hard on a string and a stream of chiles came crashing down. I learned my lesson when Estella pronounced: "*Dile pinche chile*," which roughly translates into "Tell it, 'bad chile,'" like "bad dog," or more literally and colorfully, "God damn chile."

During the day, it is women who are present in the colonias; the geography of colonia communities is the geography of the women's daily lives. The processes of immigrating to the United States, seeking documentation, creating a new personal community, and making a living all play out in the practices and beliefs of their everyday lives, just as the practices of making ristras and gossiping about community news and events are central to women's daily life. These are the processes that construct leadership in the colonias. With both the men and the women, their daily geographies dictate who they spend time with, what they have time to do, and how they share responsibilities in their homes and community. By "daily geography," I mean the set of spaces and places through which colonia residents move daily. A careful examination of one's daily geographies can map out the spaces one produces and reproduces, for example, or the spaces from which one is partitioned or excluded. These spaces tell stories that say a great deal about the relationships and power structures in which one is involved.

Moving from the particular historical and geographical juncture at which colonias were produced through neoliberal discourses of self-reliant communities, I will reveal that colonia leaders are produced as neoliberal political subjects through processes that combine the particular circumstances of colonia development with prevalent gender systems in the colonias. I also argue that colonia leaders most often come from households with more egalitarian gender systems and that this distinction is central to the construction of leadership in the colonias. In particular, households headed by single women, where there is little or no male influence, appear to be at the center of the processes that produce women as leaders. Four out of eight women in my sample come from households headed by women.

The predominance of women as leaders in the colonias is so pronounced that it almost appears to be too obvious to bother studying. Women simply appear to do their jobs as caretakers and supply the necessary resources in any way they can. But making a simple or straightforward causal argument about women's activism would be a mistake. Gender relations in the home are a very important factor in women's activism. Yet, just as the cultural aspects of colonias alone do not fully explain the production of colonias, the existence of traditional gender systems that dispose women to be leaders forms only one part of a layered process that entails the production of colonia leaders as political subjects. Traditional Mexican gender roles of women as caretakers and men as providers shape the production of political subjects in the colonias, but the role of nongovernmental organizations (NGOs) in soliciting women as leaders and the discourses NGOs produce complicate gender systems, making it impossible to read gender roles as a simple result of family dynamics. The men and women who live in the colonias operate in a transnational gender system that is a place-specific mix of what they describe as "traditionally patriarchal Mexican systems" and what appear to be more egalitarian North American systems that are affected by outside influences such as NGOs.

The now-familiar set of processes, including global economic restructuring, transnational immigration, and neoliberal governance, that led to the production of the colonias in the first place work side by side with shifting gender systems in the colonias to produce a small number of colonia leaders, the vast majority of whom are women. The same processes that contributed to the production of the colonias create the circumstances in which women experience, need, and organize for change. Thus, the dominant gender systems and structuring processes together shape colonia residents' daily practices and the meanings they attach to these practices in such a way as to construct and gender leadership in the colonias so that primarily women are attracted to leadership positions.

There is also another important element at work in the construction of leadership in the colonias: the influence of the NGOs that work within them. For many colonia residents and leaders in particular, NGOs often serve as their first introduction to multiple aspects of U.S. culture, politics, and economics, thus shaping their initial and often-subsequent impressions of U.S. society. While colonia women are at home during the day caring for their households and families, NGOs can contact them easily. This easy access makes women a popular target population for the often-very-necessary interventions NGOs make in communities. Yet, these very same NGOs are implicated in the goals of both capitalism and the neoliberal political project as revealed in Jennifer Wolch's "shadow state" and Miranda Joseph's "supplementary relationship" to community (Joseph 2002; Wolch 1989). These relationships can be difficult to navigate. It is not easy to recognize the connections NGOs make between communities and the state or the forms of governmentality and discipline in which NGOs take part. These relationships are not always easy to see because they are subtle and designed to solicit consent for the neoliberal hegemony. Many of these NGOs aim to create collective action via empowerment and politicization, activities that are far from the goals of the neoliberal political project within which they end up working. In this chapter, I will examine the role NGOs play in shaping women's entry into leadership in the colonias. In particular, I will focus on the discourses used to recruit women into leadership and how these discourses produce a specific relationship between leaders and the state that positions women leaders in such a way as to promote the values of neoliberalism.

The Production of Subjectivities

In an examination of the production of women as colonia leaders, I focus here on the ways women, through daily practices, both emerge out of and contribute to the gender relations and processes they experience. The women leaders that lie at the center of this research are central actors in the production of their community's transnational gender systems and roles. Their constant negotiation of multiple gender systems and roles they experience daily is crucial to their own production as colonia leaders. The transnational character and constant negotiation of gender roles in the colonias make it difficult to pin down exactly what produces leaders, but it also opens a large window of opportunity for gender-based changes in power relations. Due to the importance of women's leadership in the colonias and the strong connections I draw between gender roles and women's activism, it becomes necessary to understand the gendered construction of leadership before the limits to activism, one focus of this book, can be examined. This chapter demonstrates how both the production of colonias as Mexican working-poor communities and the production of women as community activists, whose work is used to serve the interests of the neoliberal political project, are linked to these limitations.

My research highlights two central and interrelated aspects in the production of leaders. The first is the many connections and similarities between the production of colonias as communities based on neoliberal values of personal responsibility and independence and the production of colonia leaders as unintentional proponents of neoliberal ideology. The second aspect to be addressed here relates to the multiple ways colonia activists view their own creation as leaders as they negotiate the many discourses available to them on their own leadership. I examine the production of colonia leaders through a discussion of both the global processes involved in the production of the colonias and the more regional processes and powers that construct a particular gendered form of activism in these communities. To complete this task, this chapter lays out an argument regarding the production of subjectivities and the ways in which subject production can be reworked by both the subjects being produced and the processes and discourses producing these subjects. This chapter also contributes to the continuing discussion and deconstruction of the naturalization of neoliberal discourses on individual responsibility and self-reliance in order to assess the limitations it places on the creation of social change in the colonias. Drawing on Aihwa Ong, Judith Butler, Chris Weedon, Allan Pred, and Michel Foucault regarding the production of gendered subjectivities, this section briefly lays out the central tenets of subject production that guide the rest of the chapter.

In the last chapter, I argued that the colonias developed as a neoliberal solution to the lack of affordable housing on the border at a particular moment. In this role, they not only house the Mexican working poor that fuel the region's economy but also produce their inhabitants as neoliberal subjects. Chris Weedon argues that "in reality individual access to subjectivity is governed by historically specific social factors and forms of power at work in a particular society" (1987, 95). Judith Butler agrees when she claims that, for the development of politically useful forms of feminist subjectivity, the "critical point of departure is *the historical present*" (1999, 8, emphasis in original).

The production of colonia women as activists and leaders takes place within specific histories. The histories and geographies of their origins determine to which discourses of femininity, motherhood, gender relations, and activism these women are introduced. In turn, the available discourses shape the kinds of activism and social change these leaders attempt: "Social relations, which are always relations of power and powerlessness between different subject positions, will determine the range of forms of subjectivity immediately open to any individual on the basis of gender, race, class, age and cultural background" (Weedon 1987, 91). As I will demonstrate, NGOs and the neoliberal state play an important part in shaping the "range of forms" available.

It is also within the particular histories and geographies in this ethnography that I see Aihwa Ong's "subject-ification" at work in the colonias. Ong's "subject-ification" is the cultural aspect of citizenship consisting of "a dual process of self-making and being made within webs of power linked to the nation-state and civil society" (1996, 738). According to Ong, to really find one's way as a cultural citizen, one must come to grips with the "various regulatory regimes in state agencies and civil society" (738). Ong is referring to Foucault's concept of governmentality and its role in regulating the conduct of citizens (Foucault 1991). In the colonias, the processes of cultural citizenship, or "subject-ification," are clearly seen in the daily lives of a set of women leaders as they negotiate their subjectivities,

subjectivities that I will demonstrate are both made for them by NGOs and the state and that they "self-make" as well. In the end, their subjectivities as leaders clearly fall into this dual process Ong describes.

While the colonias are produced as impoverished and isolated Mexican working-poor communities, the production of residents takes place in similar ways. An understanding of how discourses and social meanings structure both the production of colonias and colonia activists is a necessary precursor to the development of more politically conscious community-level change in the colonias. How colonia residents understand their communities and position in society "determines the type of society" that they "will find possible, appropriate or desirable" (Weedon 1987, 91). It is important to note that discourses themselves are practices and that the situated material practices of daily life both affect and are affected by discourses. So, for colonia women, the available discourses on gender roles, needs, and the role of the state shape both how these women develop as leaders and what they accomplish with their leadership. One important set of these discourses is produced by the NGOs that work in the colonias. As one of the central conduits for information and training on issues pertaining to resources, activism, and leadership in the colonias, NGOs exert a great deal of influence over leaders and what leaders believe is possible with their leadership and activism. NGOs are, in essence, directly implicated in the processes of activism and leadership in the colonias.

It appears that women who become leaders associate different meanings with the daily practices they share with other colonia women. I focus here on the importance of these differences. The subjectivities available are determined, in part, by the practices of daily life on which discourses of appropriate behavior are constructed. In his study of racism in Sweden, Allan Pred presses on his reader the importance of "locally situated practices" in the processes of racialization and subject creation: "It is through locally situated practices that they experience disjunctures or incompatibilities of behavior and meaning, disjuntures between the common sense behaviors and meanings of those who racialize them and the taken-for-granted behaviors and meanings they have either brought from elsewhere or have produced in syncretic fashion locally" (2000, 20). Pred discusses how in Sweden those who are racialized, the "migrants, refugees and minorities" (19), realize through their daily practices that they are part of these processes. This mix of the "common sense behaviors and meanings" of the racializers and the taken-for-granted meanings brought from elsewhere becomes the transnational space and meaning set that is the colonias. It is in this transnational space and at this intersection of practice and meaning that I argue that some women are pushed to leadership and others are not.

It is first necessary to see that there are other possible subjectivities before comprehending how one is being produced as any one particular subject. For this realization to occur, the processes that naturalize the existing dominant discourses must be disrupted and reworked (Butler 1999). In relation to feminism and feminist identity, Butler points to the task at hand as the need to create a "critique of the categories of identity that contemporary juridical structures engender, naturalize, and immobilize" (8). In the colonias, these dominant discourses range from commonsense solutions for resource deprivation to more traditional ideas on gender relations and accepted gender roles in the household. Because "individuals are both the site and subjects of discursive struggle for their identity" (Weedon 1987, 97), there is always room for resistance and reformulation of subjectivities. The key, according to Weedon, Pred, and Butler, is the development of an awareness of which discourses and practices are and are not available, why, and how the availability of discourses shapes one's subjectivity and identity. Only then is a more resistant sort of reproduction of subjectivity possible. NGOs play a central role in shaping access to discourses and practices and, in particular, in the acceptance of new discourses on resources.

While my focus is on the ways in which leadership is constructed, and in turn how women are produced as neoliberal community leaders in the colonias, I will also discuss how women activists and outsiders describe women's activism and, in particular, what they believe motivates women to become active in their colonias. Though motivations for activism and the processes that produce women as activists can be very different phenomena, they are often closely tied. The production of subjects in the Foucauldian sense occurs when dominant discourses, usually produced by people or institutions with power, construct frameworks in which certain practices, meanings, and attitudes are acceptable for some people and not others. In this way, the production of subjects who are limited in what they can and cannot do is based on the available discourses governing behavior. Foucault was more concerned with how certain institutions, governments, and family structures created the discourses that led to subject positions than he was with the subjects in these positions (Foucault 1970; 1972). For this reason, his approach to subject formation is useful when examining the role of NGOs in the colonias because their position in the development of leadership is so important.

As institutions playing a central role in the subject formation of leaders, the role of NGOs is the focus of this chapter. I have previously examined the production of the colonias as communities based on neoliberal values of personal responsibility and independence alongside their production as culturally familiar communities. These two subjectivities help to explain the poverty and resource deprivation experienced in the colonias as familiar and expected yet easily improved through the efforts of individual residents. In order for leaders to be produced in such a way as to politically challenge the neoliberal naturalization of their poorly developed communities, they must be motivated to question the discourses that naturalize their resource deprivation.

If leaders' actions are dictated, in part, by their production as leaders, how does motivation fit in? Motivations exist on multiple scales. While one colonia resident may be motivated to work on a community wastewater system to raise the resale value of her lot, another may be motivated to do so because she feels compelled to give back to her community for purely altruistic reasons. But regardless of the scale at which motivation takes place, motivations emerge from the same particular set of experiences and discourses that produce the person who has the motivation. How the production of colonia leaders takes place is closely related to the motivations leaders develop for their activism. In order to understand the limited nature of activism in the colonias, it is necessary to understand how motivations relate to the production of subjects.

I begin this discussion with an examination of the hybrid gender system that develops in the transnational space of the colonias. This gender system maintains many attributes of the traditional[1] Mexican gender systems the women experienced before migrating to the United States yet also has many influences from U.S. culture as well. Just as there is no single gender system in the colonias, as each household negotiates gender relations through their own situated daily practices in different ways, there is no single "Mexican" or "American" gender system. I do not intend to create a binary relationship between the two. Yet, the discourses produced in many colonia homes push conversations on gender roles in this direction, and my discussions in this chapter may appear to reflect this.

Family in All Its Forms

Shortly before I arrived in New Mexico, an incident happened that, when related to me at a later date, introduced me to the complicated ways in which domestic gender relations affect women's activism in the colonias. The Community-Organizing Group (COG) held its board meeting on the third Monday of every month. The board was made up primarily of colonia residents—two from each colonia with which the organization was working. There were also a few local progressive types on the board at any given time; it was their job to offer their expertise and aid in more traditional administrative aspects of nonprofit board work. While I was working with them, the non–colonia members on the board were a New Mexico State University professor and a nun from the Sisters of Mercy order. It is usually the women leaders who represented their colonias on the board. The meeting took place in town at the Catholic diocese building. Since the diocese lies just about halfway between the northernmost and southernmost colonias served by the COG, most leaders drove about the same distance to the meeting.

Transportation to the board meetings can cause problems. Most colonia households have at least one car, but that vehicle may not be running or might be needed by another family member. In order to solve the transportation dilemma, many leaders carpool. On the Monday in question, Eduardo and his wife Juana planned to drive and give Estella a ride. Everyone at the COG was pleased with this solution because all too often residents miss the board meetings and important decisions requiring a quorum must be postponed. But late that afternoon, Estella called the COG director, Elena, to say she would not be able to attend the meeting. Elena later told me that she was very surprised because she had understood that all the transportation problems had been resolved. When she asked Estella what happened to her plans to ride down with Juana and Eduardo, Estella was quiet. That was not going to work, Estella explained. Juana was not going to the meeting. Elena asked if Eduardo was still planning to attend, and Estella said that, yes, he was. It was at this point that the director realized the awkwardness in the conversation came from the juxtaposition of two conflicting gender systems: hers and Estella's. Estella then explained to Elena that, as a married woman, she could not drive for an hour with someone else's husband. It did not matter that Eduardo and Estella's husband were close friends or that she and Juana were best

friends. It was simply inappropriate. Here, the inseparability and variability of practice and meaning are clear. For Estella, riding in a car with a man
who is not her husband was very inappropriate, while, for Elena, it did not
even merit a second thought. The powerful ways in which meanings are
connected to practices kept Estella away from the board meeting and confused Elena.

The conflicts between gender relations in the home and the demands of
leadership pose a constant struggle for colonia leaders. Yet, it is these very
same gender relations and roles in the home that dispose a few women to
the activism that in turn creates these struggles.[2] The struggle these women
experience daily to improve their positions in domestic power relations is
a theme that recurred in my interviews. Family dynamics rely, for the most
part, on a traditional distribution of power. Men and women have distinct
roles and responsibilities, and it is within this discursive framework that
colonia leaders must negotiate a space for their activism.

Tradition

This section discusses the discourses of gender most commonly used
in the colonias. These are the discourses I observed in my daily interactions and recorded in my interviews; they are the discourses that shape
residents' daily lives. It is important to note here that my research focused
on the women leaders, and all my formal interviews were with women;
therefore, the discourses presented here are based on how these women
deploy these discourses about themselves and their partners. I gathered
my knowledge of the discourses specifically used by men to discuss gender
relations through observation and from interviews with the women leaders. While these discourses are mutually constitutive and created side by
side, there is a gendered nature to their deployment, as men and women
call on them in different ways. I argue here that it is these discourses of
gender that construct and gender leadership in ways that dispose women
to activism more readily than men.

Before I begin the discussion of the discourses surrounding gender
in the colonias, I want to position this discussion and, in particular, my
choice of words. I previously made reference to a "traditional" distribution
of power in many colonia households, and it is the use of this term that I
want to address now. I derived the term from my interview and observa-
tional data. "Traditional" was the word most often used by the women to

describe gender relations in their homes, and it was also the best fit for the stories they told me about gender roles in their homes and families. All the women spoke of stereotypical patriarchal gender roles and relationships to some extent. In the sections to follow, the split in daily geographies that takes men out of the community during the day to complete their role as "provider," while women remain to accomplish their caretaking work, does fit this "traditional" description. Although the reality of gender roles I encountered was never as discreet as the traditional patriarchal roles the women described in our interviews, these stereotypical and essentialized views are important because it is on these stereotypes that discourses of leadership are built. When I discuss "traditional" gender roles in the colonia household, I am referring to customary relations as described by the women activists themselves in their own, often essentialized, interpretations of gender roles in their homes.

These discourses may appear rather simplistic, but I present them as they developed in my interview data. The women in my sample acknowledged the simplicity of the discourses, and they realize their actions often contradict these discourses. Yet, they insisted this is how men and women relate to each other in general. And it is on these discourses that they based their interpretations of their daily life practices, that is, in their roles as women, wives, mothers, community residents, and leaders.

A Father's Role

In order to better understand the domestic gender dynamics I observed in the colonias, I dedicated one of my formal interviews to the topic of family dynamics in general and gender roles in particular.[3] These questions supplied the data for my observations and arguments regarding the importance of daily geographies in the gendering of leadership.

I found that, while the men are gone at work, the women are left in the colonias to care for their families and community. All the women I interviewed broke down domestic gender roles in similar ways. Fathers are disciplinarians and providers, and mothers are caretakers. Mothers take care of the day-to-day details, and, when there is trouble, the father steps in and asserts some discipline. All the answers I recorded about the family and its structure revolved around this division of family responsibilities. In most families with which I worked, when a father was present, he acted as the disciplinarian. As Juana describes, "I think the role of the father is the one who has the most authority. I mean to say, they obey him more because he has the most authority, and he is stronger because he is a man."[4] In her home, Estella liked to be the "good guy"; she much preferred her husband to do the punishing. Estella's children meant the world to her, and she did everything for them. She saved the money she made working odd jobs to give them spending money because her husband did not think they needed spending money. It is interesting to note that, even in families where no father figure existed, single mothers still relied on their ex-husbands to adopt an authority role when discipline was necessary. Some single moms called their ex-husbands and asked them to intervene when the children became unmanageable. It was not uncommon to listen to a women describe the trouble her son or daughter was getting into at school and how she was hoping her ex-husband might be able to talk, or scare, some sense into the children during their next visit with him. Although these men may no longer take part in all the traditional daily practices of fatherhood, some meanings remained intact, and they were still considered the primary authority figure in the family.

When not acting as the authority figure, fathers completed their other major role in the family, that of provider: "I know if you were to ask this question to my mom, the role of the dad is the provider, and that's basically it. And maybe they would tell you that he's the disciplinarian or something like that, but to me, I mean the role is provider."[5] Only three of the nine women worked outside the home regularly. Of the other six, two worked part time when they needed the money or a job interested them. Yet, all the women viewed their partners as the primary earners in the household. Perhaps this is because the men are full-time earners, while the women took time off to have children, care for grandchildren, or attend to the home. Women also viewed supporting their husbands in their work as an important element of their role in household gender relations. For Marie, the father is the family's "foundation." In this analogy you can see the mutual support and dependence many women feel toward their partners: "I think the role of the papa in the family is basically to give security to the family. [To give] support, economic support . . . the strength, he is the strongest in the house. Let's suppose that a house needs a well-made foundation. I think the father is that foundation."[6]

For Estella, providing this support meant spending extended periods of time on the road with her husband, who runs a national trucking company from their trailer. When I met Estella she was in her late thirties; I never did get her to admit her true age. All that mattered to Estella was that she was older than I was, and therefore I had no business pressing her for details regarding her age. Like many colonia women, Estella had a large extended family in the area. As we know, her parents and a cousin lived on the lot next door; her brother Mario and several other siblings also lived nearby. She also had a grown daughter with a family of her own who lived about an hour away. This large support network allowed Estella to travel with her husband without worrying about her children or home. She usually took her five-year-old daughter with her and left her sixteen- and twelve-year-old sons in the care of their grandparents. Estella viewed her travel with Jose Miguel, her husband, as a vacation of sorts. She did not have to prepare meals or do the housekeeping, and she could spend the day talking and laughing with Jose Miguel. Her daughter thought her father's semi truck was the "coolest house on wheels." But just picking up and leaving for days at a time can hinder a leader's organizing activities. Projects had to be put on hold while Estella was gone, and she often felt overwhelmed when she returned. Yet, she believed it was her job to help her husband in any way possible, even if this meant sacrificing her own work as an activist. Without Jose Miguel's income from trucking, they would not be able to afford Estella's sport-utility vehicle or the supplies the children needed for school.

Estella was born in Mexico and grew up in Juarez, where her father settled the family so he could be closer to the United States where he worked as a bracero in the agricultural fields of Texas, New Mexico, Colorado, and Illinois. As the second oldest in a family of six children, Estella started to work later than most of the other women I interviewed. She worked from sixteen to twenty-one in maquiladoras in Juarez. Before she went to work, she married for the first time at fourteen and had a baby. This marriage lasted a year, and she returned to her family and her work in Juarez. In the early 1980s, Estella's father managed to get a work permit that allowed his family to live with him in the United States. While there working in the fields, Estella met her second husband, Jose Miguel. Born in El Paso, Jose Miguel was a U.S. citizen, which made acquiring citizenship easier for Estella than for most of the other women in my sample. In 2000, they had been married nearly twenty years. Jose ran his own trucking business, owned two semis, and employed at least one driver full time. When Estella was not busy working on community business, she helped him with the invoices and truck schedules. She also made money on the side doing some seasonal work in the fields and at home. Estella was one of the best-known colonia activists in the area, which means that NGOs often called on her to consult on projects that affected the colonias. Through these contacts she received short-term work on projects in a number of fields, including community health and daycare provision.

A Mother's Role

As mentioned earlier, the most common role for a mother expressed to me in my interviews and daily experiences was that of caretaker. Mothers make sure their families are fed and dressed and ready for another day. These practices define motherhood in the colonias. Several women spoke of their role as that of an educator. Marie described the mother as the one who has the intelligence and wisdom in the family: "Well, I say that the role of the mother is even greater than that of the papa, because if the mother does not fulfill her responsibility as a mother, well everything falls apart . . . the woman is the intelligence and the wisdom. I believe that without intelligence and without wisdom I don't know how it is possible to survive as a family. Of course, with God's will, because without God's will we are nothing, right?"[7]

Juana described why mothers must be educators and authoritarians as well: "Well if the mother is too lenient with the kids, the kids do with one whatever they want. One has to also be strong of character. Because otherwise the kids are not well educated. And now if they are not well educated with what you tell them, well, that's their problem. Because one is teaching them to get ahead so they will be better in life, better than you are. And if they do not want to get ahead that's their story."[8] Without the proper discipline, children will take advantage of their parents and end up "mal educado." This literally means poorly educated, but is more commonly used to mean troublesome, misbehaved, and ignorant of what is right. Juana believes mothers must teach their children how to get ahead in life and how to have better lives than their mothers. The theme of mothers helping children to achieve better lives recurred throughout my interviews with women activists.

Flora expressed quite eloquently the hardships of single motherhood: "Well everything [is the role of the mother], wash [the kids' clothes], iron [the kids' clothes], educate them, give them food, talk with them everyday, get them ready for school, and aside from that one also has to work, so it is hard."[9] Not only must Flora fulfill all the caretaking responsibilities of a mother, but she must also take on the traditionally male role of primary wage earner. Yet, even with all this time-consuming work, she acknowledges the importance of talking with her children every day. At times her house was less than spotless, but Flora would rather take the time to be sure her children were on the right track. On several occasions, I watched Flora take time out to talk to her eleven-year-old daughter about school and life, especially boys.

Discussions between mothers and children are an important way to observe gender roles and relations in action. Many of my impressions about gender roles and gendered power relations came from being present during parent and child conversations, lectures, and arguments. These conversations elucidated the pressing issue of sexual activity and teenagers. Teenage, and even preteen, pregnancy is a big problem in many communities, and the colonias are no different. Some of the girls in these communities have babies at age eleven. Each month, I saw a new preteen mother proudly carrying her baby around to visit the neighbors. For many of the preteen mothers I encountered, their first pregnancy actually occurred before their first menstrual cycles. Without their periods to alert them to the possibility of getting pregnant, they often did not realize they were already fertile. The young women I met do have a basic idea of how human reproduction functions, but they are vague on many of the most important details of their own fertility. This lack of knowledge is reinforced by the absence of sex education in the schools and the social taboos that exist against open discussions of sexuality in their culture.

I had the opportunity to sit in on a workshop given by a local parenting resource center on the topic of sexuality, peer pressure, and young adults. Through this workshop, I learned not only a great deal about the women in my study and how these mothers pass down ideas of sexuality to their daughters but also how the NGO that ran the workshop was involved in the complicated processes through which ideas about sexuality move between Mexico and the United States and settle into the transnational space of the colonias. The workshop took place once a week in a different home in the colonia of Recuerdos for a month. All the participants were women, though men were welcome. This was the one of the few times I felt comfortable to speak freely with these women and girls (two of the participants were preteen mothers) about their sexual experiences. In most of my day-to-day experiences the sexual histories of the women in my sample, and their ideas on how to educate their daughters and sons about the facts of life, did not come up in conversation. So, this program was a great opportunity to engage in topics we had not discussed in detail before. The program was intended to encourage parents to help their children with the many pressures youth experience. The workshop facilitators found the lack of sex education in this community deeply problematic. The children of Recuerdos go to school in Hatch, where a highly conservative school board had ruled against offering sex education at both the middle and high school levels. What limited knowledge these young women had about their sexuality came from their mothers, older sisters, girlfriends, and, often, older boyfriends. At the time it did not occur to me that terminating sex education programs in public schools is not just symptomatic of a religiously conservative town; it can also be tied to neoliberal policy. The neoliberal political project has had very controversial and visible effects on public education in the United States, particularly at the elementary and high school levels. In the name of shrinking "big state" policies that include sex education in public school curriculum, the neoliberal political project sees sex education as something better left up to the family and personal values. The No Child Left Behind Program is perhaps the best known education project that clearly exemplifies the market and competition-based logics of neoliberalism as applied to education via a testing-based system (Hursh 2007). When the neoliberal state fails to offer a necessary service, in this case sex education, NGOs step in and fill this gap.

The workshop began by asking the participants to explain how they learned about their own sexuality. To my surprise, several of the women said they had no idea as girls that they would ever be menstruating and their first periods caught them by surprise, often shocking them. One woman said she did not know for years what was happening to her each month, and she kept it a secret from her entire family. Another said that, after several months, she figured it out and that with time and her first pregnancy, the rest of the human reproductive cycle became clearer to her. The girls in the group appeared to know a bit more, but not enough to avoid their first unplanned pregnancy, a fate they all agreed they would have avoided if they could have. All the participants agreed that more information was needed and that they had tried to offer more support to their children than they had been provided. The limiting factors appeared to be unfamiliarity and a lack of comfort discussing sexuality based on a set of cultural discourses that construct it as taboo and an inappropriate topic for casual discussion.

The underlying theme appeared to be avoidance. There was a subtle fear that if sexuality were discussed openly, children would be more likely to be sexually active. These attitudes toward sexuality, mixed with the lack of sex education in the schools and the prevalence of sexual activity in teen culture, clearly resulted in the minds of these women, both young and old, in preteen pregnancy. There is a fascinating similarity to be found between the aversion the conservative village of Hatch had for sex education in their schools, for moral reasons, and the discomfort many Catholic Mexican families have with discussing sexuality. This similarity, of course, was the reason for this course in the first place and was a focus of many discussions.

This workshop offered new and alternative discourses of sexuality that directly challenged the preexisting and dominant discourses. When exposed to these alternative discourses, the colonia leaders and other women in the workshop all wished to engage in these new ways of thinking about and approaching the sexuality of young adults. It is in moments like this, when old discourses and practices are brought into tension with new discourses, that colonia residents and leaders affect their own production as subjects. In this case, these new and alternative discourses encouraged the women to produce themselves and their children in new ways as sexual subjects, and it was an NGO that offered these women the alternative discourses of sexuality.

"A better life for my children" was one of the primary motivations for activism that the leaders mentioned, and one of the main ways they worked for this better life was to educate their children to avoid the mistakes they had made. The women's desire to help their children lead better lives makes the topic of teen pregnancy an important family conversation for many of the leaders with whom I worked. Flora felt very strongly that her daughter should wait to have a family and live a little first; at the very least she wanted her daughter to get a good education. In order to make this point, she took Sara, her daughter, to visit a friend who had been missing from school for a few months. This girlfriend was now a full-time mother and did not have time to finish the sixth grade. At first Sara thought the baby was great and was impressed with how mature her friend appeared, but by the end of the visit it was all too clear to Sara that her friend had lost her freedom and had grown up too quickly. Flora's plan worked perfectly. When Sara described her friend's life to me, the picture she painted was far from perfect. This sort of hands-on education in the daily difficulties and responsibilities all children experience, but colonia children experience even more, was one of the roles in which Flora took the most pride.[10]

Women as Heads of Households

I rarely did interviews while men were home. Leaders and nonleaders alike usually invited me over while their spouses were at work, mowing the lawn, running errands, or out with their friends. When men were around, they usually excused themselves to watch television or go outside. Where the men were was never an issue in the homes of the four leaders who lived in women-headed households. Of these, Flora and Esperanza were divorced single mothers, Josephina was a widow, and Rosa lived apart from her husband, who worked in California.

Leaders from households headed by women and those from more traditional patriarchal households take part in similar daily practices. Yet the meanings associated with these practices differ. For leaders from women-headed households, buying groceries and school supplies are more than just caretaking activities. They are practices that embody the differences between their homes in which they are in charge and those in which their colleagues are, at best, a partner in a household, at worst, something akin to a servant. Without men to play the traditionally male role of provider, leaders who run their own households must be caretakers and providers and these added roles bring new meanings to daily practices.

For leaders like Flora and Rosa, the lack of men in their household also meant more freedom daily. Neither woman had to be home to prepare an evening meal for their husbands. Unlike the cohabiting women in my sample, Flora and Rosa's daily lives were not based on the schedules of others to such a great extent. They did need to prepare lunches for school-age children and be sure there was food for dinner, but they could plan meals for times that fit their schedules, and they could change plans as they pleased. The other women in my sample often described their lives as organized around the demands of their families, especially their husbands who expected them to be available at all times to prepare food or run errands.

Single mothers also experience activism in different ways than their married counterparts. Flora, the heart of the organizing movement in Los Montes, was the first person to whom outside organizers, NGOs, and the county went when they needed help. Flora is an extreme example of the competing responsibilities colonia leaders must balance. As a single mother, she juggled both the traditional male role of provider and the domestic responsibilities of a mother and caretaker. The two single mothers in my sample, Flora and Esperanza, had both more responsibility and more freedom. Without a regular male presence in their homes, Flora and Esperanza created a different gender system than those in the other leaders' homes. Single mothers offer interesting new gender systems in the colonias that fuse and challenge the male–female role dichotomy. Single-mother activists, by taking on both male and female roles, can be seen as self-sufficient and self-motivated neoliberal superstars—able to single-handedly run both home and community.

Flora was born in Chihuahua, Mexico, in 1966. As a child she worked in the fields with her parents when necessary, but mostly she lived a quiet rural life. At fifteen, she left home to travel around Mexico and work. At sixteen, she came to the United States for the first time and worked watching children and cleaning houses. This was a difficult time for Flora, as she experienced financial hardship and unfair employers. When she was twenty-two, a cousin introduced her to her husband, and they married two years later. The marriage lasted six years and produced two children. While I was in Los Montes, her ex-husband still lived in the area with his parents, and they were in regular contact. Before they moved to Los Montes, Flora and her husband lived with his parents in a nearby town. But after a few years of this living situation, Flora decided she wanted a place of her own, so they bought a lot in Los Montes. During my time in Los Montes, Flora was thirty-four and lived with her two children, a twelve-year-old daughter and a seven-year-old son. Flora took classes on and off to get her general education degree or GED (i.e., the equivalent of a high school diploma) because her formal education ended in Mexico in her sixth year of school. As a single mother, she worked as much of the year as she could. Since I have known her, she worked at an airstrip, in the pecan plant shelling nuts, in a chile plant making chile powder, and at McDonalds. Since many of Flora's jobs were physically strenuous, once or twice a year Flora took a month off just to regain her strength. It was often during these breaks that she would become most involved in community organizing. These breaks were thus invaluable to her leadership efforts. Flora described herself as strong and motivated, and she believed she had a lot to give to her community. Like many of the women in my sample, Flora named her children as one of the most important motivations she had for her activism.

Transnational Nature of Gender Roles in the Colonias

The transnational nature of the colonias permeates every aspect of daily life. Sometimes this mixture of cultures runs smoothly through daily practices, and, at other times, the pressures of negotiating a culture that spans two nations are too great, and contradictions and tensions abound. Nowhere in all my research was this more pronounced than in household gender relations. In general, and rather stereotypical terms, all the women described similarly gendered roles for men and women in the home. The gender roles described above lay out a set of rather traditional patriarchal roles that all the women agreed were more or less present in their homes and the homes of their neighbors. Yet many of these women are active daily in leadership roles that blur the traditional boundaries between gender roles. Although they describe men as the providers, these women provide their colonias with some of the most important community resources, such as potable water and paved roads. I argue that close attention to the transnational nature of gender roles and the roles of NGOs in the development and maintenance of traditional gender roles in the colonias can shed light on the many contradictions and tensions women experience between their roles as mothers and leaders.

The transnational mix of gender systems also manifests itself in the variety of gender relations and roles seen in colonia families. Although the women all described similar roles for men and women in the home, each woman's own life and family displayed variations on these traditional roles. For example, Josephina lived part of the year in Los Montes and the other part in a small mountain village in Chihuahua. She took her first long-term trip to the United States shortly after her daughter Flora's first child was born. Since then, she returned regularly whenever Flora needed help with her children. Flora and her children were not the only attractions for Josefina in Los Montes, as she also had three other daughters and eight more grandchildren spread around the colonia. Only her youngest daughter remained in Mexico. Aside from the obvious physical, transnational nature of her family, Josephina also had to learn to negotiate the less-obvious meanings in each of her daughter's homes, where slightly different transnational gender systems were at work. Grandparents and great grandparents often move back to Mexico, or commute as Josephina does, once they have saved some money and are ready to retire. Her trips back to Mexico gave Josephina a much-valued break. Yet even with all her responsibilities helping to care for her grandchildren, Josephina still found time to lead her adopted community.

Josephina nearly always could be counted on to be around, which made her an invaluable resource for organizing. In 2000, Josephina was close to sixty years old and she did not like to talk about her age. The youngest of nine, she was born in the rural village of Santa Rosalia, Chihuahua, around 1940. After four years of school in Mexico, at the age of eight she started to work in the fields and in the house to help her family make ends meet. Her father was a cowboy while her mother stayed at home. At the age of sixteen, she left her family and went to Chihuahua City to work. When she was twenty-one, she married a cowboy whom she met while working in the fields, and they remained married for nineteen years. In Mexico, they had five daughters and one son.

The transnational nature of Josephina's extended family is very common in the colonias. Josephina was only involved in her colonia's leadership with her daughter Flora, a single mother. Her other three daughters in Los Montes were all married and had chosen not to be active in their colonia because they did not believe their husbands would approve. Flora once commented to me on the high percentage of single mothers in colonia organizing: "The activists are ninety-five to ninety-nine percent pure women. Single women because those that have husbands, the husbands won't let them [be active]."[11] When I asked her why this was, her answer was simple and very telling: "Because men are *machistas*, they want their women here in the house with them, they want hot food on the table."[12] But her analysis of the situation did not end there; she had obviously given it a lot more thought: "It's also a women's fault, she gets used to [living under her husband's control]."[13] Gender roles and gender systems develop differently in colonia households based on the presence or absence of male heads of household.

Juana, a leader in Recuerdos and a longtime friend of Estella, was the only woman in my sample who actively participated in her colonia's leadership with her husband, Eduardo. Though slightly more involved than her husband, Juana preferred to work with Eduardo whenever she could. Gender roles in her household mirrored those in the other male-headed sample households, even though Juana worked outside the home more than the other women did. Juana was born in Mexico, in Guanajuato, and lived there until she was twelve, when she moved across the border to El Paso to work. She worked alone and was often sad during the four years she cleaned houses until she met and married Eduardo, who was also born in Mexico, at age sixteen. Once married, they moved to Hatch, where she met Estella. Unlike Estella, Juana worked nearly full time during the agricultural season. She worked beside her husband and older children and only left the fields early in order to go home and prepare dinner. The work Juana did every day was backbreaking. She came home reeking of onions or stained red from chile, fearful to touch her eyes because the chile oil on her hands would make them burn and water for hours. Yet when asked if she would prefer to do something else, Juana said no. She liked the freedom of the fields and the fact that she could work outdoors with her husband and children. They talked and teased each other and still got work done. Juana liked to describe her and Eduardo as friends more than husband and wife, a description few of the other women ever used. She also liked the money. Juana and Eduardo were two of the fastest field workers I saw. If the crops were good, they made nearly enough money to last the whole year so they could take it easy during the off-season. Although always a bit shy when the tape recorder was going, when I shut it off, Juana shined. She was outspoken and very insightful. She has taken a lot of time in her life to think about her own experiences and how she would like things to differ for her children. Juana described herself as a woman of strong character, a woman who got things done, especially in regards to her children.

Esperanza was different. She had a difficult life and, at times, her vision of the future was not as clear. Getting from day-to-day was a challenge for her. Esperanza was born in Durango in 1959, the fourth of eleven children. Her father was a laborer who worked in the cotton fields, mines, and sheep pastures. Her mother stayed home and cared for the children. After completing three years of school in Mexico, at age eleven she left her parent's house to work. When she was twenty-five she came to the United States with one of her brothers to work and, while cleaning houses in New Mexico, met her husband, who was from Chihuahua. They married and stayed together for fourteen years. Esperanza divorced before I met her and lived with an older son from a previous relationship and two sons from her marriage. The boys all lived with their mother in Los Montes, while their father lived a few miles away. She separated from her ex-husband because he suffers from several serious psychiatric conditions and never got the kind of help he needed to control his violent behavior. Esperanza, like many colonia women, spent a few weeks in a women's shelter while she tried to get her husband out of the house.

The transnational gender systems produced in colonia communities can mean more freedom for women activists, but they can also signal more gender-based discrimination and abuse. As immigrant's lives develop to sustain daily geographies that transcend borders, this can mean the creation of domestic gender systems that are a hybrid mix of those they brought with them and those they have recently encountered. This mix can be more or less egalitarian, depending on many factors such as education, employment, and socioeconomic status, among others. In the colonias, many women discussed what they saw as an opening for greater equality in the home as they reached for opportunities to move beyond Mexican patriarchal machismo and its ever present role in their relationships with their husbands, fathers, and sometimes even sons. No matter what the nature of the gender system that develops in a particular household, domestic violence can be a common problem in the colonias. Traditional gender relations that stress patriarchal power relations and privacy make it hard for friends, neighbors, and even family to intervene. Immigrant women often fear they will be deported if they report the abuse. While I was in New Mexico, I helped the local diocese and the bishop draft a pastoral letter on domestic violence.[14] Though the priests and I often differed on wording, I contributed ideas based on the input I gathered from the women in the colonias. The women were pleased that the church was addressing the issue, but they were also embarrassed that the bishop and others were aware of the violence in their lives. Several women told me how important it was for the church to speak out against violence. They were upset because in their past experiences it had been priests who told them to stay with their abusive partners because marriage was sacred and divorce was not an option.

Many undocumented immigrant women in the United States remain trapped in violent relationships. The bishop's letter on domestic violence was in part an effort to support the 1994 Violence Against Women Act,[15] which grants undocumented women the right to press domestic violence charges against their abusers without jeopardizing their status in the United States. Many of the women I met in the colonias had no idea they could safely leave their abusive partners and expressed doubt at the existence of such a law. Most of these women had been convinced by their abusers that their legal status in the United States was conditional on remaining in their marriages and relationships with their abusive partners.

Negotiating Transnational Gender Relations:
When the Men Are Gone

One of the clearest themes to emerge from my interviews was that of men's greater involvement in the "outside" world. In their daily practices as providers, men leave the community every morning and return in the evening with their wages. Once they return from work they have completed their responsibilities for the day. According to Sylvia, "Men work 8:00 to 5:00, 9:00 to 5:00, whatever, come home, eat dinner, maybe water the trees, and go to bed. That's their role in the community."[16] Once a man is home, he is done for the day. Anything else they choose to do once they are home in the evening "is purely recreational" and unpredictable. Most men do not choose to take on extra responsibilities, and therefore work done in the community commonly falls to the women. As Mario made clear, "For the men it's like 'OK, I'm going to work from 7:00 to 5:00, and if I work for that time, I've accomplished my role."[17]

Most of the women I interviewed believed their partners had little desire to do anything once they got home, recreational or not. Husbands usually did not want to plan or attend the community meetings necessary to make change in the colonias. Because many colonia residents work during the day, be it in the fields, factories, or in their homes watching their children, community meetings must take place in the evening in order to allow the greatest number of people to attend. Marie described her husband when he gets home from work in these terms: "Men are the head of the family and it's more comfortable to say, 'Vieja [old lady] you take care of all of this, don't bother me; I'm watching baseball, I'm watching the news, I don't want you to bother me.' And then it is the woman who has to take the initiative, make the decision to do what she believes is best to do, even if later the viejo [old man] gets mad."[18] The same discourses that dictate that men provide incomes generated outside the colonias shape the scales at which the daily geographies of men and women become manifest. The scaling of daily life for colonia residents is closely tied to, even constituted by, gender discourses. In her discussion of the distinctions between public and private spaces, Linda McDowell acknowledges the construction of this dichotomy: "The division between the public and private, just like the distinction between geographical scales, is a socially constructed and gendered division" (1999, 149). In keeping with McDowell's argument, I would like to add that the construction of the public–private split, geographical scales, and gender discourses are all mutually constitutive. The practices and meanings that shape the scales of the discourses on which colonia leadership is based are interconnected in many ways. It is the dominant ideology of gender and gender roles and the labor market that dictate that men are the providers who must leave the scale of the community and enter the scale of the regional economy in order to be able to provide appropriately. Women, by contrast, play out most of their daily life at the community scale. As with most gendered social relations, the scaling of daily life in colonias also follows the public–private split. The community is associated with the private domestic sphere, while places outside the community, where men work, are tied to the public sphere. In this way, colonia men align themselves with the public sphere, and leave the private sphere of daily life responsibility in the home to the women. Yet, in the process of working to improve the conditions in which they must care for their families and community, women leaders cross over the public–private split and become political subjects in the public sphere of county, state, and federal politics.

The combination of public and private spheres and community activism illustrate a complex relationship. For the most part, the women view their work as community and private sphere based. Most women use discourses of community and personal need to describe their motivations for activism. But, they do acknowledge that the organizing can take them to more public scales. I agree with McDowell's description of the theoretical muddiness of this situation, as she wants to "blur the sharp association between gender and space and suggest that there is a messier and more complicated set of relationships to be uncovered since so many activities transgress the clear associations between femininity and privacy on the one hand, and masculinity and public spaces on the other" (150). The activities these leaders engage in daily rewrite both what their communities believe to be acceptable for women and often what they themselves feel comfortable doing. Yet, as I will demonstrate, these very well-worn dichotomies are set at the foundation of the discourses used by both NGOs and state agencies as they recruit women with whom they partner in the colonias.

The daily lives of the Mexican women leaders with whom I worked clearly demonstrate this "messy" complexity. Any work that moves one's personal life into the public sphere in an intimate way, as community activism does, must be a complicated process. The practices of daily life take on new meanings as they move out of the colonias, and colonia leaders must adapt to these differences. One of the many difficult elements of community activism and leadership is the negotiation of relationships and roles that cross the public–private divide.

With men at work all day, the women have more or less free run of their households and, to a lesser extent, the colonia. In many colonia households, women act independently of men during the day. Women leaders find it easier to be active in more "public" areas than they would if their husbands were present. I routinely scheduled interviews for working hours when husbands could not listen in on or show disapproval for my interview questions and their wives' answers. As an outsider and researcher asking often quite private questions, I acted as a sort of lightening rod for issues about academic research, community organizing, and gender relations in the home. Many men clearly felt uncomfortable with my role in their homes. They accepted their wives' activism as long as it appeared more necessary than political and did not disrupt existing gender relations. The relationships between gender, power and practice that our conversations raised for these women appeared to be a bit too political and challenging for some husbands. For example, none of Flora's sisters felt comfortable talking to me in the presence of their husbands, who already questioned Flora's leadership role. They told me their husbands thought they were busy enough just taking care of the household and did not need any more responsibilities, especially ones that might hinder their domestic duties. Perhaps Flora's insight into the marital status of activists came from observations of her own family and its marital complexities.

Marie was not a stranger to marital struggle around her leadership activities, nor was she shy about letting anyone present at community meetings know about her husband's discontent with her activism and her involvement with colonia business. Ernie and I arrived in Los Montes and Valle de Vacas at about the same time, so, while I was getting to know the colonias and their leaders, he was initiating a recruitment campaign. The meetings and workshops that made up this campaign served as a great introduction to both the COG and how they viewed the role of other NGOs in the colonias, and most importantly, the relationships between colonias and their leaders. It was in this series of meetings that I first heard Marie explain her worries about the future of community organizing in Valle while she and Rosa were waiting at Rosa's house for more people to arrive at a meeting. They were discussing why more people did not attend these meetings, even though she knew everyone in the community was aware of them. It was not uncommon for activists to hand out flyers to every house in the colonia and then only have one to five percent turn out to a meeting. In this small percentage you would usually have a few couples but mostly women by themselves, either women who live without men in their homes or those who simply turned out that evening without their husbands. If any member of the couples retuned, it was often just the wife. Discouragement was an all too common feeling. At this meeting, Marie was starting down a very familiar road. She wanted to make it clear that she could not do all the work on her own. She was "president" of the community, a title she loved to use, but her husband Marco, did not want her to do it any more. Marie was also frustrated because she did not speak English and this was a problem "because the county speaks English."[19]

Ernie piped in right away and pointed out that the county ought to provide a translator so language should not be a barrier. The marital issue was a bit touchier. We all knew Marco and that he could be a pain. We also knew he was easier than many other husbands; after all, Marie was a leader. He may bother her, but he does let her out. While he was at work during this meeting, Marie was happy because, as she said, "I don't need to be checking my watch to see if he's waiting for me."[20] We were all thinking the same thing when she said this; he could be a lot worse, and this was the very reason there are so few women at this meeting.

When the men have attitudes like this, it is not surprising that women do most of the community organizing: "Because the man thinks that with work and bringing in the money he's done his part," explained Estella, "The woman, no, we are always thinking, what's better for the kids."[21] Here Estella pointed to the ways in which the discourses men use define their roles and limit their activity in the family and community. Estella believes it is the women who think about the future of the community the most, particularly in relation to the children.

The simultaneous production of women as leaders in two spheres of daily life, the domestic and the community, reinforces their production as political subjects who support the interests of the neoliberal political project. As discourses surrounding gender roles are engaged, they are used to encourage women to take part in activism at the same time that they tie women to the community and private sphere, both foci of neoliberal policy. Neoliberal policy focuses on these areas in order to support the development of grassroots movements that will pick up the slack in areas where the state pulls back from resource and service provision. In the colonias, we see this quite clearly in the work of women activists who agitate at the intersection of traditional gender roles and neoliberal ideology to create change in their communities.

The relationship between men's lack of motivation to work in their colonias and their family responsibilities is a complicated one. In her work on the environmental justice movement and women's participation, Celene Krauss found that men saw their wives' activism as threatening to their role as providers: "Toxic waste issues thus set the stage for tremendous conflict between these women and their husbands. Men saw their roles as providers threatened: the homes they had bought may have become valueless . . . they were asked by their wives to take on housework and childcare. Meanwhile, their wives' public activities increasingly challenged traditional views of gender roles" (1998, 145).

By protesting the environmental hazards in their communities, the women in Krauss's study called into question the provisions (in the form of housing) these men had made for their families. For many colonia families, these infrastructure-lacking lots are all they can afford. As the region's favored pool of cheap labor, Mexican immigrants barely make enough to meet their needs. In a culture where provision is a primary aspect of the traditional role of a husband, it is only reasonable that men might believe their wives are pointing to a flaw in their ability to provide by focusing so much attention on what their communities lack.

In turn, husbands uncomfortable with the meanings and repercussions of their wives' activism also lay blame on the NGOs that work in their communities by organizing their wives and offering services. For many husbands, these NGOs represent everything they find threatening about their wives' activism, especially because they do this work in a professional, organized, and very vocal manner. This may be why many husbands discourage their wives from taking part in NGO activism and why even the husbands of activists stay away from NGO organizers and are sometimes less than friendly toward them.

One such husband was Marco, Marie's partner, who often gave NGO staff members a hard time and made their work that much more challenging. Marco was polite and was always a gentleman; he would never challenge a female NGO worker the way he would go after a male NGO staffer. Similarly, he only sat in the background and observed my interviews, never really interrupted, but only made his presence known. As

Marie and I would sit at her small kitchen table, I could often see Marco out of the corner of my eye lying on the bed watching television in the next room. He always had the volume low so he could hear us, the door was never shut all the way, and he often made excuses to come into the kitchen. We both knew he was checking up on us.

But when I went to Marie's, the situation was completely different with Ernie. Valle de Vacas was one of the colonias Ernie organized, so he had quite a bit of experience with Marco, and it was a running joke around the office that Ernie and Marco did not get along. Because Marco worked the swing shift, it was very likely that Ernie would run into him when he went by to visit Marie. During these encounters, Marco would play devil's advocate and drill Ernie on all the organizing efforts of the community and why they were not working faster, getting better results, or making more changes. Inevitably, Ernie had to explain that organizing was a slow process and one could not expect change overnight, and Marie would agree. She would point to all the good work she and the community had accomplished. Most of the time she would busy herself with getting some coffee ready for Ernie and me, and when Marco had said his mind he would wander back to the television or for a nap before work. This scene would repeat itself again the next time Ernie and Marco met. It became predictable, and Ernie grew to dread running into Marco.

The irony, of course, is that Marco was becoming aware of some of the real issues at the heart of the relationship between NGOs and colonia communities. The meshing of neoliberal discourses on self-reliance and independence with traditional discourses on gender relations produce leaders, primarily women, who not only keep their communities' problems, and solutions to these problems, internal to the community but also create solutions that follow a traditionally gendered division of labor and leave traditional meanings associated with gendered practices intact. NGOs play an important role in the construction and gendering of leadership in this situation, as they seek out women who fit their preconceived notions of leadership. NGOs also construct activism as another form of mothering when they link the work leaders do to their daily activities as caretakers. These connections help NGOs recruit leaders and naturalize leadership activities that might otherwise intimidate some women. NGOs also use self-help to make leadership and activism seem more accessible. For example, when the colonia leaders decided they needed wastewater systems, the neoliberal discourses that direct both their communities and the agencies that work in these communities pointed toward self-help methods, such as digging trenches and building their own system, as the most feasible solutions. The use of self-help meant the leaders would not need to examine more political solutions, such as seeking legal recourse on the part of the county against the developer, because their colonias could just do the necessary work themselves, thus keeping the women within the community and out of the local public sphere. Self-help rhetoric, in turn, produces women's activism and its results as less like politics and more like good caretaking, and the women appear less like the "activists" who are currently under fire from conservatives across the globe, not to mention some activist's husbands in their own kitchens.

Doing What Comes Naturally: How Women's Daily Responsibilities Prepare Them for Community Organizing

All the women in my study are mothers and the primary caretakers for their children and grandchildren. These women identify as mothers and wives above all else. This description may sound limiting; to label a woman as a "mother above all else" would be considered by many to be reductionist. But the women with whom I worked often described themselves in these terms. Their daily lives revolve often around their roles and responsibilities as mothers and wives, and it is in these experiences that they are frozen as they are produced as leaders. It is in their daily practices as caretakers that women have the most firsthand experience of the resource deprivation that characterizes the colonias. When these experiences of need meet the neoliberal rhetoric found both at the foundation of colonia development and in most state and private interventions in these communities, women leaders are produced as political subjects who serve the interests of the neoliberal political project.

The Role of NGOs in the Neoliberal Production of Colonia Communities

Neoliberal ideologies of self-reliance and individual attainment in the place of collective action are not cultivated by women leaders alone. NGOs play an important role in this process as well. In her book, *Against the Romance of Community*, Miranda Joseph uncovers the pivotal role "community" plays in sustaining capitalism through a legitimating of the social hierarchies on which capitalism depends (2002). In the colonias, "community" is used to support neoliberalism through the efforts of women leaders. According to Joseph, it is "precisely through being cast as its opposite [that] community functions in complicity with 'society' enabling capitalism and the liberal state" (2). As colonia residents rally around their community and its improvement, they take on the cause of community organizing, and with that they come into the sphere of influence of a growing number of NGOs. These NGOs, in turn, are implicated in the construction of leadership as serving neoliberal interests. NGOs can be seen as agents of the neoliberal political project as they implement programs that reflect the project's stress on personal responsibility and individual advancement at the cost of any form of greater or collective good. NGOs in the colonias are implicated in the processes of neoliberal governmentality. One of the ways NGOs exert their influence is in recruiting leaders. NGOs use their knowledge of gender systems and daily practices in the colonias in order to locate women they believe fit their model of the colonia leader. In this way, NGOs construct not only the shape of leadership but also the entry points to leadership.

How Family Roles Dispose Women to Activism

Discourses in the colonias dictate that women's domestic responsibilities lead them into and prepare them for activism in their colonias. It is commonly believed that many of the skills women leaders use in their organizing and leadership were learned through their roles as mothers and wives. Several women told me if it had not been for their experiences dealing with bureaucracy in their children's schools and in health care clinics, then they would never have believed they could tackle the county's bureaucracy. As primary caretakers, these women had no choice but to learn to deal with large government institutions to get their children's citizenship paperwork, Social Security cards, and access to programs like Women, Infants, and Children. All these jobs may seem routine, but to a recent immigrant or a woman who does not speak English and who cannot read, these tasks are nearly insurmountable. In addition, it is often the networks women develop as mothers that allowed them the flexibility to complete their work (Stack 1975). The daily practices of motherhood and leadership are similar enough that some women can make the move from one role to the other.

Raising a family and running a household is a full-time job for many colonia leaders, yet they find the time to organize and improve their communities. Many leaders and NGOs believe colonia activists can run both a household and a community because these tasks draw on a similar skill set. The connection between the skills learned in the home and those used to organize a community is acknowledged by outside organizers and used to recruit and maintain women leaders. By linking the discourses of basic need provision and neoliberal self-reliance, state and private agencies, through their programs and rhetoric, participate in the production of colonia leaders as activists who serve neoliberal policy agendas. This is done, of course, through "community" organizing, thus reflecting Miranda Joseph's theory of the mutual dependency between community and capital and, in this case, neoliberalism as a political project grounded in a capitalism that serves the interests of the wealthy.

COG organizers, who realize the connections between running a household and running a community, actively promote these similarities when recruiting women. Mario had been a community organizer in the colonias of Doña Ana County for more than ten years. As an outside organizer, Mario was not a community leader. Although he was a colonia resident, he did his organizing in colonias other than the one in which he lived, and he was paid to do this work. Mario knew most of the women who led the colonias and worked with them on many projects. He experienced women's leadership on three levels: as a colonia resident himself, as a community organizer who recruits and trains leaders, and as the brother of an important leader.[22] Mario acknowledged the importance of daily life experience for colonia leaders. He believed women's experiences as mothers and problem solvers in their homes enabled them to take on the role of leaders in their colonias: "For the women, they are prepared to deal with everything, and I think that is why they spend the time on the community. Say, if I have raised five children, this [leadership], this is nothing. Going to a meeting is nothing for me."[23] Mario drew on a rather stereotypical idea of motherhood in supporting his argument, and it is important to note that this essentialization accords how the women identify themselves. Raising children is all about planning. According to Mario, so too is community organizing, a similarity that he recognized. The valuable experience that Mario and the women leaders believed motherhood instills in women is crucial to understanding why women take on leadership roles in the colonias. These women know if they can organize their households, then they can organize their communities by working with other women who have gained similar leadership potential from their experiences as mothers (Gilmore 1999).

Activists such as Sylvia agree with Mario's observations.[24] As a colonia activist herself, Sylvia understood why women, and not men, assume leadership positions. She was one of the women who rose to the occasion. When asked to explain why women take the lead in colonia organizing, Sylvia said, "I think it comes from just running a household, and now they've got to start running the community, too, because otherwise nothing's going to get done."[25] In her opinion, the skills acquired running a household, such as managing and organizing money and time, better enable women to run community organizations (Kaplan 1997, 94).

Even women who do not choose to become leaders are produced as conduits for information and contacts for outside interventions. Colonia populations can be very unstable. People come and go as migrant field labor, and others return to Mexico for extended periods. Because of the nature of the contract-for-deed documents through which colonia residents buy their land, serious gaps often exist in county records regarding colonias. For this reason, if the county, or anyone for that matter, wants to circulate information in the colonias, they must use one of two methods: they can go door-to-door, or they can call on the schools to send flyers home with the students. This second method works well because nearly every colonia household contains at least one school-age child, and the children usually give these flyers to their mothers, not to their fathers, so the message ends up in the right place. Sylvia perfectly described the importance of what seems like an everyday after-school scene: "Kids come home from school with a little paper that an organization left, and they say 'Look, mom,' they don't say 'Look, dad.' They don't give it to the dad, they give it to the mom. And the kids are learning so much in school and educating the parents, but the first one to get the education is always the mom. It's very rare that a child will go home and say to his dad, 'Look at what the teacher gave us today.' It's usually the mother, so I think that is why they tend to get more involved."[26] The sharing of knowledge and education between children and mothers is very important to the dynamics of development in colonia communities. This is one of the ways in which the production of children as subjects of neoliberal policy takes place. It is very common for mothers to take children to NGO-sponsored meetings and workshops because they cannot afford childcare. Many NGOs provide childcare knowing this will entice more women to attend. Although children are playing or otherwise entertained during most of these meetings, they still become acquainted with the activist and NGO culture and are often aware at an early age of the topics that are important in their communities. Several NGOs, COG included, had special organizing sections that focused on young-adult issues in the colonias and in this way recruited activists at younger and younger ages.

The very same experience that aids women in their community organizing also burdens them. Naples writes of the "triple day" many women perform as a result of the racialized and gendered division of labor (Naples 1998a). As working poor Mexican women, the women in my study were expected, within a gendered domestic division of labor on the scale of the household, to be the primary caretakers in the home. In the racialized division of labor at the regional scale, they were only offered low-wage employment. When the women did find paid employment, they encountered a triple burden; they were family caretakers, low-paid workers, and unpaid community activists simultaneously. Community activists found it hard to balance these three roles. Often the perfect balance was impossible to accomplish, and some aspect of their triple burden fell aside. Even for those that balance only two of the three daily tasks, the demands on their time may be overwhelming. The women in my study found that if one aspect of their position needed to be sacrificed, then they abandoned their community work for other, more pressing issues. If they were working outside the home, they could not sacrifice that income. Likewise, they would never think of neglecting their domestic duties, especially those involving their children. But even when they did fall behind in their community activism, the women would always return to it and pick up where they left off. It is this sort of dedication of which Nancy Naples writes when she contends that it is the unpaid work of women community workers and activists that "sustains the social fabric that constitutes viable communities" (191).

Necessity: The Mother of All Invention and Change

As colonia leaders learn organizing skills through their daily household practices, they also experience a high level of resource deprivation. A community's lack of running water most often affects the women. Women must clean the house, do the laundry, prepare meals, and take care of the children—tasks that require water. When it becomes necessary to haul water from a central spigot or from the water truck, these tasks become both very time consuming and arduous as they take on new meanings. The lack of resources with which colonia women must deal turn mundane practices, like doing the laundry, into constant reminders of the need and resource deprivation their communities experience. For the women who become leaders, these constant reminders push them to activism. Several of the women in my study remember a time when their colonias lacked running water, and daily chores were much more burdensome. What is most interesting, though, is that women did not use these memories to explain why they became activists. Rather, women presented them as a measure of the improvements they had already made. Instead, they discussed the roads that were paved by their work and the parks they designed. This minor distinction in the language the women used to describe their activism is very telling. The women and the NGOs with which they work construct leadership as something that comes naturally from women's role as caretakers. Therefore, when it is difficult to complete this role, women are pushed to action.

As caretakers of their homes and communities, colonia leaders become active in their communities on multiple scales. The resource deprivation and need that help shape the production of women as colonia leaders are experienced in a variety of ways. Most commonly, women experienced their communities lack of infrastructure at the household level when they cleaned and cooked. However, they also experience this absence at the scale of the body itself through their situated daily practices, such as bathing or using the toilet. Each leader has his or her own set of personal motivations. These motivations contribute in important ways to the women's political development as leaders.

Rosa articulated the motivations for her activism in explicitly personal terms. She was the oldest of eight children born to a farming family in Zacatecas, Mexico. Her father worked in the fields in Mexico and in the United States as a bracero. At nineteen, Rosa left her parent's house for California and the chance to travel in the United States. In 1975, at age twenty-eight, she married and settled in Norwalk in Southern California. Rosa, her husband, and their two sons and daughter moved to Valle in 1994 to take part in a dairy business her brother started. Once they arrived, the business fell apart. After her husband built their house, he moved back to California where he could work in construction and make more money. It is not unusual for colonia residents to move in and out of the state using networks of family and friends to secure better or more consistent work. Rosa lived with her two youngest children, and twice a year her husband came out to visit. Rosa said she is involved in the colonia because she wanted to create a park for her physically disabled son. It was very difficult to move him about, and if there were a park in the colonia, then it would be much easier for her to take him for walks and get him outside.

"My own sense is the reason women end up doing it [activism] is they're kind of the backbone."[27] This is the response John gave me when we first met to discuss his long-standing work in the colonias. John is an old-time activist. He got his start doing medical work in Central America. When I was working in the colonias he ran the Border Water Group (BWG), a nonprofit that helps colonia communities put in self-help wastewater systems. Because it does not meet health and environmental standards, true self-help wastewater technology is not legal in the United States (Carew 2001). What we call "self-help" in terms of wastewater in the colonias is simply basic, resident-installed septic systems. Truly self-help technology based on renewable resources, mulching systems, or other alternative technologies used successfully in the global south are not legal in the United States. Thus, many communities go without any sort of wastewater systems for decades, since they cannot afford to install systems that meet federal standards. When money could be found to help pay for wastewater systems, John's organization brought the equipment and knowledge to help colonia residents do a lot of the expensive and backbreaking manual labor themselves. With John's backhoes and earthmovers, colonia residents dig all their own trenches and lay the pipe themselves. In this way, their own sweat puts equity into their land and they do not have to pay others to do the work. Without this type of sweat equity, the projects would never get off the ground. When I met John, he had been doing this work for more than a decade, and he knew colonia residents all over Texas and New Mexico.

John saw how the women's daily experiences of need pushed women into activism: "They're the ones that have to deal with the kids not having water or not having water to bathe with," he said. "I think women are somehow more affected by these daily life issues, so they're the ones that said, 'We're going to do something about this.'"[28] Though men do most of the physical labor that John's self-help wastewater projects involve, women are the ones who come to the organizing meetings and make sure the colonia residents are there to do the work.

While I was in New Mexico, work was done to lay the lines for a wastewater system in the colonia of Valle de Vacas. Marie single-handedly did almost all the organizing necessary to get this project under way. Marie worked hand in hand with John and his on-site man, Jesus. The process began in 1996, when the colonia of Valle de Vacas, at the direction of the NGOs, COG, and BWG, applied for a grant to install the lines for a wastewater system. They first had to prove to the county that they were serious about the project. To show their enthusiasm for the project, a group of colonia residents went to the county commissioners meeting. Marie believed the community received the money because they were united: "In this petition it said that if they gave us the money for the drainage system, the community would put in the manual labor because this would advance the project faster."[29] When the colonia finally heard that the county would give them the grant, they were overjoyed. The news came at a community meeting in the colonia of Los Montes, which was also included in the grant. Marie related with pride that the district's county commissioner was at the meeting himself to present the good news. Valle de Vacas and Los Montes received $400,000 for the implementation of a wastewater system, but the hard part was still to come.

As the key organizer on the project, Marie was in a bit of an odd situation. As a woman, she was not expected to do any of the manual labor, yet as a community leader, she was expected to make sure the colonia supplied the necessary labor daily. Marie had to create an acceptable gendered division of labor. Within this gendered division of labor, the men did the manual labor and the women scheduled the project and provided lunches and snacks. When the organizing got too difficult (many families said they never thought the lines would reach their houses so they refused to help), Marie told me she would motivate herself to keep going by remembering when she had been deprived and how hard life was without the most basic infrastructure. She takes courage and motivation from the needs of others in her community. She articulated this courage in classic neoliberal ideas that valorize individual responsibility and self-reliance as the key to community development.

Sylvia agrees that women in the colonias experienced daily need more harshly than the men: "Women have to drive the streets day in and day out, take the kids to school. They're more involved as far as water goes because they're the ones that use it more than anything, they're the ones that wash, the ones that cook with it. They're involved with the gas more because they're the ones that are cooking and knowing if they're running out of gas or not."[30] Sylvia points here to women's involvement with building infrastructure daily. While most North Americans consider basic resources commonplace, in colonias, resources like running water and electricity are neither common nor in place. Esperanza expresses a similar view as Sylvia: "Well one needs the services more, right? The man, not really. The man, he finishes work, comes home from work, eats and baths, and that's it. We need to do more, to take care of the kids. We meet and get the services."[31] Esperanza is keenly aware that both men and women use the resources, but only women appear willing to do the work necessary to procure them.

The lack of infrastructure in many colonia communities encourages some women to take initiative and make changes in their communities. Colonia residents know they are living resource-deprived lives and consider it unfair. Yet as colonia leaders are constructed, in part by NGOs, to facilitate relationships between colonia communities and the neoliberal political project, they are produced as subjects who provide for themselves rather than question why they are without in the first place. What takes place in the colonias surrounding the production of leaders is a nexus of discourses from those relating to the production of transnational gender systems, to U.S. discourses on Mexican labor, to colonia discourses on leadership, to NGO discourses, often created with funders, on women's caring work.

The Importance of Outside Contact

Women's daily geographies make them more present and available during the day. Thus, they become the target of most outside interventions. For example, Josephina and Marie share similar daily patterns but in different colonias. Both women watch their grandchildren during the day, saving their children a great deal of money. Josephina spends most of the year living with her daughter Flora. Without Josephina's help, Flora could not survive. As a single working mother, Flora needs help getting her kids ready for school in the morning and caring for them when they return home. Marie's grandchildren come to her house while their parents work. Both Josephina and Marie start their days long before the children

are awake. While the rest of the house sleeps, these women prepare breakfast for the household and lunches for those who go off to work or school.
Once Marie's grandchildren eat and Josephina's grandchildren depart for
school, the women begin the household work. First, they do the morning dishes and eat the leftovers, their own breakfast. After the morning
meal is out of the way, they do a few loads of washing. All the women
with whom I worked had washing machines in their homes. When you
have a large family and do agricultural or another kind of manual labor
for a living, investing in a washer is a wise choice. A dryer is not as necessary as the dry desert air does the job nicely. By late morning, the women
finish most of the household chores. Then it might be time to go to the
market or see the doctor in the local clinic. Once a month on average,
an afternoon must be spent in a county office dealing with immigration
paperwork, work visas for visiting relatives, or any number of other situations recent immigrants encounter.

At least a couple of afternoons a week, these women receive visitors.
The most frequent visitors are relatives, many of whom live in the same
colonia or a nearby community. Aunts, cousins, nieces, and nephews stop
by to enjoy a cup of coffee, some pan dulce, and the local gossip. When the
visitors are not relations, they are most likely to be other colonia residents
seeking information on colonia projects or advice on personal matters.
The women leaders with whom I worked are considered to be experienced and wise women in their colonias.

Once a week or more, community organizers and county workers visit
colonia leaders. Many outside organizations that hope to gain access to
the colonias approach these women. On many occasions, I accompanied
an organizer from the COG on his or her daily rounds of the colonias.
Although they live in different colonias, both Josefina and Marie work
with the same COG organizer, Ernie. He often visits both colonias on the
same day to discuss similar projects, and the women know each other.
Their meetings with Ernie usually involve a cup of coffee, a few cookies, an
update on colonia news and gossip, and finally, an hour or so discussion of
colonia organizing business.

One afternoon I spent with Ernie in Los Montes with Josephina and
Flora we walked around the colonia with a county map trying to identify
which family lived where and how many trailers existed in the community
as a whole. This should have taken only an hour, as there are fewer than
forty lots in the colonia. But we quickly found that the county's idea of

where a lot began and ended greatly differed from the lot owner's idea. In several cases, we simply could not identify the lots as the county presented them, and we had to draw our own map to clarify the county's obviously outdated version.

On another afternoon in Valle de Vacas, Marie, Ernie, and I spent several hours discussing how to use Marie's position on the board of a local water company to the benefit of the colonia. Sometimes during these meetings with Ernie, the women would sit at the table with him for hours and just talk. Other times they would politely excuse themselves from the table and keep on working on the household chores that remained. Though Ernie would offer to help, the traditional gender roles the women employed made it unacceptable to let him help with domestic chores. I, on the other hand, helped wash many dishes and vastly improved my ironing skills.

Women leaders provide the contact point for most local and federal interventions in the colonias. Many of the community groups that exist in the colonias were created in conjunction with outside organizing groups. Many, such as the COG, are nonprofit organizations that help colonia residents create community groups through leadership workshops and organizational assistance. Outside groups can create a level of excitement about community organizing that residents alone have trouble duplicating. For example, the COG has archives of information about infrastructure building and other topics that interest colonia residents. The COG also has strong contacts with county officials and can help facilitate meetings between colonia leadership and the county. Although they hold many meetings in the evenings, the bulk of the COG's organizing work happens during the day. COG organizers have families and lives outside their work, and they prefer, like most people, to spend their evenings at home with their families pursuing their personal interests.

NGOs and Leadership

It is necessary to examine the role of the COG and similar outside organizing groups, particularly in relation to the construction and gendering of leadership. The COG plays an important role in the construction and gendering of leadership in the colonias. During my time working in the colonias, I worked side by side with the COG on many occasions. As a nonprofit that strives to create community organizations in the colonias, the COG has a vested interest in colonia leadership. One of its primary functions is to identify and foster potential leaders. In this function, the COG recognizes the greater freedom of many single women and women who live away from their husbands to take part in leadership activities. On several occasions, Ernie and Mario commented to me that leaders like Flora had more room in their lives to pursue their activism. Yet, these women's time was also demanded by the greater responsibilities put on them in the absence of men in their households. In its recruitment and retention of colonia leaders, the COG both takes part in the production of political subjects that serve the interests of neoliberalism in the colonias and encourages women to question the legitimacy and efficacy of neoliberal discourses. This contradiction characterizes much of the work NGOs do in the colonias.

COG organizers first enter a colonia and locate leaders and activists by word of mouth. Once they make contact with the existing leaders, they help them in their organizing and recruit more leaders to help with the work. Approximately one third of my study population were existing leaders when the COG entered their colonias. The COG and the existing leadership recruited the other two thirds. The conclusions reached in this chapter on the factors that construct and gender leadership and those that produce leaders in the colonias apply both to women recruited by the COG and those who came to organizing by other means. Although "outside contact" may have a slightly clearer effect on the recruited leaders, existing leaders had contact with outside influences as well. For example, existing leaders had experience with county officials, such as housing inspectors and environmentalists. They also had experience working with the New Mexico Department of Health and local health clinics that come to the colonias to publicize their services.

The leaders in my sample came to their activism and their relationship with the COG in different ways. Marie's family history of activism motivated her to start her work as a colonia leader. COG organizers saw Marie's inclination toward leadership and used her family history to guide her further into her activism.

Marie was the ninth child in a family of thirteen children. She was born in 1952 in Mexico and lived there until she was nineteen, when she came to the United States for work. She met Marco, her husband, in the United States and has been here ever since. When I met Marie, of her four children, her youngest was twenty-one and still lived at home. For Marie, activism was a family legacy handed down to her by her father, a Mexican labor activist and union leader. In our later interviews, Marie mentioned her father on several occasions and even pulled out some of his old union papers, which she kept safely in a drawer, both as a memento of her father and as inspiration when she got discouraged. Although her husband exerted a great deal of control in the house, Marie never seemed intimidated and always made time for, and took great pride in, her activism. While researching media attention on the colonias, I found several newspaper articles with interviews and photos of Marie. Clearly, she had nothing to hide when it came to her community leadership activities. As I got to know her, I did come to realize the pressure Marco exerted on her activism and his constant disapproval of her leadership activities. This weighed very heavily on her mind. Although she appeared 100 percent dedicated to her cause, she was plagued with worries about her ability to continue on as a leader.

While Marie had to balance her desire to serve her community with her need to please her spouse, Alicia had no such issues. She simply had little free time to dedicate to activism. Alicia lived around the corner from Rosa on a street with only three trailers. She often complained that even if she did get all her neighbors involved, only three more people would be active. But as it was, she only spoke to one set of neighbors who were already as involved as they wished to be. Most of the time they just relied on Alicia to acquire the information they needed. One of the key roles for any community leader is that of information provider. As Alicia became better known in Valle de Vacas as a leader, she was called upon more often to provide information to the colonia. But Alicia was not like most women in the colonias; she had a college education, and she taught high school Spanish in El Paso. Unlike the other women, Alicia spoke fluent English. At twenty-nine, she was also the youngest woman in my sample group. She and her husband were born and raised in the United States. Yet, she came from Mexican roots and had a lifestyle just like the women who were born in Mexico. Both of her parents, born in Durango, had only a couple of years of schooling each. Alicia stayed with her parents in El Paso until she and her husband married at age twenty-three after seven years of dating. Alicia and her husband were both employed full time in El Paso, had more job security, and made better money than many colonia residents do. They chose a colonia lot because it offered them the most for their investment and the privacy for which they were looking. Working full time and raising two small children took up most of Alicia's time, yet she still tried to find time for her colonia. As one of the only women leaders in Valle de Vacas with a working computer and fluency in English, Alicia often made the flyers for meetings and typed up notes and letters to the county.

Of all the women I met, Alicia articulated the reasons for her activism at the most structural level, regularly referencing issues of inequality. Other women acknowledged that the conditions in which they lived must have some sort of political basis, but they focused their analysis of the situation on need and their desire for immediate results. Alicia often spoke of the discrimination colonia residents encountered and of the rights she felt she had as a U.S. citizen, which were being ignored, such as the right to a certain standard of living. The COG organizers actively encouraged Alicia to take a bigger role in the leadership of Valle de Vacas, but Alicia was very good at saying "no" when she lacked the time. In her own words, Alicia was a very "strong-willed" woman with a bit of a temper and enjoyed being an individual who made her own decisions.

Once women are contacted by outside organizations and become active in their colonias, their availability during the day continues to be important. Many of the most pressing activities in which community leaders take part happen during the day. For example, in order to reserve a room in the local school for a community meeting, a leader needs to make the reservation during school hours. When the room is reserved, a community leader must then contact the county experts she wants to have speak to the community at the meeting and this, too, must occur during business hours. If a leader had a nine-to-five job and could not access a phone to make these calls while at work, meetings would never get scheduled. Finally, on the day of the meeting, hours and hours must be spent preparing the food that will be served at the meeting. An informal meal, usually a one-pot dish like menudo or chile, followed many of the meetings I attended. The women leaders prepare the meals for two reasons. First, food is always a good incentive for attendance. Second, the time spent socializing after the meeting is very important for maintaining a sense of belonging, especially for those who are not part of the community organizing daily.

Sherry Cable's work on women's participation in an environmental protest organization points to the practical level on which "homemakers" have schedules more appropriate for activism than their husbands who work during the day: "The data suggests that it was practical necessity.

Many SMO [social movement organizations] tasks required weekday performance, when husbands were working. Thus, structural availability led homemakers to assume duties sometimes beyond traditional gender roles. Successful experiences encouraged similar steps in other areas, expanding their repertories of appropriate gender behavior" (1992, 46). Cable also makes the connection, as have I, that women's activism both comes from their "traditional gender roles" and yet transforms the very same gender roles. Although differing in race, class, and geographical location, the women in Cable's study fell into activism in ways similar to those in my study. Being in the right place at the right time is an important element of women's activism both in Cable's study and in the colonias. Both Cable's data and my own point to the importance of women's community networks for recruitment and retention. After initial involvement in the SMO, women's availability during the day makes them better suited for social activism than their husbands.

Taking this attention to women's "traditional gender roles" a step further, NGOs in the colonias intentionally target women as they recruit potential leaders and activists. As Mario and Sylvia's words attest, the language of caretaking is central to the subject formation of colonia leaders and activists. As Mario explained previously, "If I had raised five children . . . leadership is nothing to me." The connection here is clear: if you can raise children, you can raise a community. At this level of analogy, this movement of experience from the private to the more public or civic realm might be considered empowering or even, if done in the right manner, politicizing. Certainly this latter result was the COG's primary goal. Yet, this is not the kind of social transformation many NGOs bring about in the colonias. Rather, many NGOs are agents of a very different process of social change, one that disables rather than enables and depoliticizes instead of politicizes.

The Partnering State

In the colonias, we see NGOs working with what has been termed the "partnering state." Wendy Larner and David Craig see neoliberalism in New Zealand as having been replaced by "a new form of joined up, inclusive governance characterized by relationships of collaboration, trust and, above all, partnership" (2005, 402). Although the move beyond the neoliberal political project has not been as clearly theorized here in the United States, and in the period in which this research took place (between 2000 and 2008) neoliberalism was certainly deeply entrenched, I did see the partnering system taking hold. This is particularly true at the grassroots level, where NGOs play a large role in the processes of "best practice" governance and where partnering programs are especially popular. This leads to "new hybrid forms of governance" that in the colonias are seen in the actions of leaders and in their choices of projects (Larner and Craig 2005, 413).

The partnering state with which colonia activists work is theorized also by Gerda Roelvink and David Craig as a masculinist state that is entirely reliant on women's labor for its partnering program to work (2005). Taking the discourses that Mario and Sylvia reference, which naturalize women's work in the community, the partnering state moves women's responsibility for caring for the home to caring for the community as well. Roelvink and Craig point out that, in many partnering programs of both advanced liberal and neoliberal regimes, women end up in "parent-friendly jobs" that actually combine mother and worker only to benefit the state: "Participation in these terms simply exacerbates the double burden" (116).

The partnering state concept is based on Wendy Brown's theory of "the man in the state" (1995). In this theory, one way in which Brown describes the masculine power of the state is in its power as a form of dominance: "This dominance expresses itself as the power to describe and run the world *and* the power of access to women; it entails both a general claim to territory and claims to, about, and against specific 'others'" (167). Partnering is a new form of access to women at the grassroots level and one for which NGOs serve as a particularly good conduit for access, especially in the colonias and other working-poor communities in which neoliberalism has left great needs unfulfilled: "The task of tidying up and joining up after all the fragmentation and conflict of neoliberal governance is falling disproportionately on feminized labour from the community and voluntary sector" (Roelvink and Craig 2005, 107).

Colonia activists work to "tidy up" their communities and help NGOs to implement state-funded resource-provision programs that create the infrastructure their colonias did not receive when they were built during the height of the neoliberal era in the early 1990s. If taking care of the community is the same as taking care of the home, then women should be naturals for the job; they should be willing, and it should not be an imposition. This is how activism is often presented by the partnering state. It is seen as a natural element of a woman's role as a caretaker, not a public and certainly not a *political* act. For many leaders' husbands, these discourses, which naturalize leadership and tie it to caretaking and domesticity, are reassuring. As I have shown in Marie's home and that of many nonleaders and would-be leaders, one of the greatest conflicts I encountered was the conflict between husbands and wives who disagreed on the amount of time activism should take up in a woman's life. If the partnering man had his way, this would never be an issue because activism would be discursively produced as such a natural offshoot of caretaking that any good mother or wife would be lining up to take part. As it is, most colonia activists live in women-headed households. It is the depoliticization of activism that is so natural in the partnering state that is dangerous.

Conclusion

Colonia activists and scholars consider it common sense that women are the leaders in the colonias. Women are in the communities day in and day out, and, in their roles as mothers and wives, they experience resource deprivation most intensely. These circumstances lead some women to take action. This chapter has complicated these very basic, yet valid, explanations of women's activism by forging necessary connections between the production of colonias as communities of Mexican working poor that support the neoliberal political project and the production of colonia leaders as independent and self-reliant political subjects. The circumstances that led to the development of the colonias play a large role, alongside transnational gender systems and the meanings leaders attach to daily practices, in the production of colonia leaders. This attention to the simultaneous production of multiple subjectivities sets this argument apart from most discussions of women's activism in the colonias.

I argue that the production of the colonias as resource-deprived Mexican communities and the production of women activists as leaders who serve the interests of neoliberalism happen simultaneously and in response to similar local- and global-scale processes, in particular, the global economic restructuring that has spurred Mexican immigration to the United States, which is then combined with the ongoing popularity of neoliberal policies and a lack of affordable housing, which led to the development of colonias in the first place.

These processes also shape the transnational gender systems and gendered power relations that, alongside outside influences such as NGO interventions, construct and gender leadership in such a way that lead women into the majority of leadership positions. In particular, women from households headed by women have more freedom to be activists and attach a different set of meanings to their daily practices as caretakers. For women who run their households on their own, daily practices, such as shopping for food or fixing a leaky faucet, are about independence and self-sufficiency as much as they are about caring for their families. As Flora said at the beginning of this chapter, "Everything is the mother's role." When single mothers like Flora take on the traditionally male role of provider alongside their roles as caretakers, they become self-reliant and independent women who "do not need men," according to Flora, and "can do it all on their own." For Flora, her daily practices that provide for her family are filled with meanings of independence, an independence the married women I met in the colonias did not mention. This variation in meanings associated with daily practices make neoliberal discourses of personal responsibility and independence more familiar for women who run their households. This different set of meanings attached to common daily practices can explain why half of all the leaders in my sample lived in women-headed households. The results and style of women-led activism are closely linked to what discourses are available to leaders and, more importantly, which discourses ring true to women leaders.

These discourses are shaped in part by the NGOs, which both construct a very particular picture of leadership for women in the colonias and then, through their workshops and interventions, produce a specific type of leader, who is shaped in multiple ways by their role in the processes of neoliberal governmentality. Often shaped by their associations with an increasingly "partnering state," many NGOs inadvertently bring neoliberal governing processes and discourses into colonia communities, and these discourses fit with much of what many leaders already know as they live very independent, self-reliant lives.

· II ·

Women and NGOs

CHAPTER 3

Empowerment and Politicization in the Colonias

Poco a poco se anda lejos.
Little by little, one goes far.

Bathing *al Fresco*

While we were walking in the midday sun, Ernie and I were doing some door-to-door visits in the colonia when we decided to see if the new woman was in. We had heard independently that there was a new young mother living in a trailer on the back of a lot in Los Montes and we thought she might like to take part in some community activities. Cold calls like this can be difficult, especially when we knew the woman in question was living in pretty rough conditions, so most likely her life was very busy. Concha was living on an illegally subdivided lot. Since it had become illegal to create new colonias in the mid-1990s, the biggest problem had become the subdivision of already-existing colonia lots by their current owners. Concha was renting a very small plot of land for her trailer, and, without any papers, she could not hook up to the water lines or electricity. She was running a power cord from the neighbor's trailer and getting water from his spigot. We had few expectations about her aspirations for a leadership role in the colonia.

As we passed the first trailer on the lot, I was taken aback by the site we encountered. In front of a very rundown trailer was Concha, leaning over a large metal tub containing two small children and steaming bubbly water. Apparently, we arrived at bath time. As we approached the tub, the kind in which you might wash a large dog or use to hold drinks at a barbeque, Ernie introduced us, and my mind raced through the mechanics of the situation. Clearly, as we had suspected, Concha had no running water and therefore no plumbing, so bathing was an outdoor affair. My first thought was: did she bathe out here as well? If so, she had no privacy at all. My next thought, how did she heat the water? She must have filled the tub with the hose we saw running from the trailer in front of her trailer. Did she possibly do this by heating the water on a hot plate or was she running a stove off an extension cord? That was unspeakably dangerous. Either way, she had to have carried pot after pot of hot water out of her trailer to fill the tub. All I could think was that Concha's daily responsibilities took infinitely longer without running water and only the power of one extension cord. My final thought was, there is no way she had the extra time to devote to community activism; she had her hands full just doing the caretaking activities required of her as a mother. She is going to think we are crazy asking her to take part in the community-organizing efforts while watching her bathe her children in a tub in her front yard. Even with the guidance of the Community-Organizing Group (COG), she could not possibly have the spare time or energy.

I was rather abruptly shaken out of my thoughts and disbelief at her situation when Concha's face lit up and she enthusiastically said, "Yes, yes, I have heard about you, and I would like to take part."[1]

Not surprisingly, Concha's enthusiasm could not live up to her circumstances, and she did not become a leader at that time, although she did attend a few meetings. From what I hear, she has maintained some interest in the community's organizing. Building wastewater systems, paving roads, and laying pipe for natural gas are not usually the everyday fare of stay-at-home mothers. In between overseeing these projects, leaders might do a load of laundry, bathe a child, or bake a birthday cake. The saying "all in a day's work" takes on new meanings for stay-at-home mothers in the colonias.

In this chapter, I examine the results of community activism in the colonias. I use the term "results" here in a very general sense; this discussion is really about what women's activist work *produces*, including desired outcomes, unintended consequences, and, sometimes, failures. In these communities, women's leadership produces physical improvements, a greater sense of community, personal growth for those involved, and new forms of governance through their involvement with nongovernmental organizations (NGOs). Some projects proceed smoothly, while others are doomed from the start. A firm understanding of what activism does and does not produce is necessary in order to analyze the limitations, contradictions, and ambiguities of women's leadership. The products of women's leadership are important because they present a concrete and material site from which to examine the daily workings of activism in the colonias. It is also the products of activism, or its results, that are often a point of contention and debate. Where to focus a colonia's limited resources is rarely a unanimous decision within the community. Once a community does come to some form of agreement on a project, they may find outside groups, like the COG, have a different agenda, and, in order to receive the help and funding on which they rely to get many projects off the ground, they must reprioritize the colonia's projects to match the NGO's priorities.

Activism and what it produces result from the situated, everyday practices of leaders, and it is through activism and leadership that women produce and reproduce the spaces of the colonias. As an important form of social production in the colonias, women's leadership can be seen as the most promising way to create social change in these poorly developed communities. Yet politically conscious, progressive change that confronts systems of power is not achieved easily alongside physical and infrastructural improvements. To increase the production of politically conscious change, I argue that it is first necessary to understand what activism and leadership in their current forms do and do not produce in the colonias. Social change is not used here to describe physical changes in the colonias but rather the "social," cultural, and intellectual changes that can go along with infrastructural change.

Social change means making changes in ideas, prejudices, and understandings of processes and power relationships that limit or oppress a group such that people can work to improve their own lives or those of others and create changes in these processes and relationships as they do so. In the colonias, the changes made to address these power differentials might be infrastructural, but they would be made based on the resident's understanding of their place in these power relationships and how the creation of new infrastructure affects and possibly confronts their position in these networks of power and oppression. The current and past infrastructural improvements were rarely made in such a manner.

I argue that leaders improve the physical environment in which they and their families live, but, due to their discursive production as neoliberal leaders of self-reliant communities and their relationships with NGOs, these activists have a hard time creating progressive social change that challenges the systems of power that marginalize them as Mexican immigrant workers. I focus in this chapter on the limitations to women's activism in this context and ask, how is it that women's leadership is both enabled and disabled in the colonias? At the center of this apparent contradiction are colonia leaders, NGOs, and the state as it serves the neoliberal political project. This three-way relationship encourages women leaders to be active and produce material change while it discourages them from following through with the more philosophical and political complements to this change.

Once women's leadership is better understood, it will be possible to propose ways to integrate progressive social change with material change in the colonias. In this and the next chapter, I argue that these two products of women's activism need not be as mutually exclusive as they currently appear to be. In order to integrate them, it is necessary to understand why they remain so distinct at this time. This chapter will identify the obstacles to enacting more encompassing progressive social change and set up the next chapter as well, which examines how these obstacles are produced.

Empowerment and Politicization

Empowerment and NGOs appear to go hand in hand. NGOs that build latrines are just as likely as NGOs that build polling places to claim empowerment as one of their contributions to community development. The widespread use of the term can be traced to what William Fischer calls, "the general sense of NGOs as doing 'good,' unencumbered and untainted by the politics of government or the greed of the market . . . NGOs are idealized as organizations through which people help others for reasons other than profit or politics" (1997, 442). Practitioners, politicians, and academics rarely question the empowering effects of these idealized NGO interventions. This clearly raises a question: what do we mean by the term "empowerment"? In a continued effort to focus on the ambiguities and contradictions present in the colonias, it is necessary to engage with the concept of empowerment–not because it is held up as the most important result of activism in these communities, but because it is "overused and abused" (Staudt, Rai, and Parpart 2001, 1252).

Empowerment literature can be roughly broken down into three categories: empowerment to control or govern, empowerment as a goal of individual attainment and happiness, and empowerment as a means to social justice and community advancement. The first grouping views empowerment as a relationship based on power. In these theories, empowerment becomes a bad word. It is something to do to another person; a person one holds power over. These theorists draw on Foucault's theories of governmentality and in particular, technologies of self and biopower to explain how empowerment can be used to control another person while seeming to give "power." Barbara Cruikshank's work on empowerment falls into this category (1999). The second typology stresses the happy side of empowerment, and these theories often can be seen as overly optimistic. Examples of this abound on international organization Web pages such as the United Nations Development Fund for Women.[2] These discourses tend to refrain from challenging ideas of empowerment in any manner, and this is why theorists such as Raymond Bryant find them troubling (2002). I believe it makes them perfect for neoliberal organizations, like the World Bank, to use to encourage *individual* women and men to empower themselves as "responsible *individuals.*" The final category places empowerment on the road to politicization. In these theories, empowerment as a personal choice is located within some kind of social change or social justice–based goal. Jo Rowlands's work is located within this typology (1997).

Wendy Brown's use of the term "empowerment" in her work on power and freedom sheds light on my arguments about the ambiguities and contradictions contained in much of the empowerment literature (1995). Brown demonstrates the ways in which many discussions of resistance employ discourses of empowerment that are neither empowering nor resistant. She notes that "contemporary discourses of empowerment too often signal an oddly adaptive and harmonious relationship with domination insofar as they locate an individual's sense of worth and capacity in the register of individual feelings, a register implicitly located on something of an otherworldly plane vis-à-vis social and political power" (22). In the colonias, we see empowerment as an individual goal as well, rather than a collective or community goal. Detachment from social and political power is also visible in the colonias where leadership and activist work focuses on improvements to the colonia in discourses framed almost entirely internally to the communities and their residents with little or no reference to the outsiders who developed the communities in the first place.

This individualization of experiences of, and responses to, power can have damaging effects. In colonias, emphasis on individual empowerment not only aligns residents with neoliberal ideology but also limits the community-wide social change women leaders can bring to their communities. The often exclusive focus of empowerment on individual emotions and personal growth allows discourses of empowerment to "converge with a regime's own needs in masking the power of the regime" (23). In the colonias, the neoliberal political project masks its goal of limiting collective empowerment through its provision of individually focused affordable housing.

An increasingly common critique of empowerment is the lack of awareness of the power relations present in empowerment situations. In her book, *The Will to Empower*, Barbara Cruikshank focuses on what she terms the "relations of empowerment" (1999). By this she means that the "powerless," or those who are going to be "empowered," do not actually exist in this state of powerlessness prior to being chosen, so to speak, by those who have "the will to empower." The relationship that develops between those who are "powerless," but will become empowered, and those who have what Cruikshank terms the "expertise" to do the empowering is of course a power relationship; thus, "relations of empowerment *are* simultaneously voluntary and coercive" (72, emphasis in original).

This may sound rather sinister, but the will to empower can be done with the best intentions, as it has been in the colonias by most NGOs. Yet, at the same time, it remains "a strategy for constituting and regulating the political subjectivities of the 'empowered'" (69). Cruikshank and others who are critical of the concept of empowerment base their critiques on a Foucaldian analysis of citizenship. Cruikshank uses Foucault to discuss the ways in which citizens are created with specific subjectivities and then empowered to be "active" subjects. Similarly, Peter Triantafillou and Mikkel Risbjerg Nielsen draw on Foucault's discussions of governmentality to demonstrate how the power relations that surround discourses of empowerment encourage the "powerless" (to use Cruikshank's term) to "constitute themselves as free and active subjects" (Triantafillou and Nielson 2001, 67). In the colonias, we see empowerment at work on two scales. NGOs work to empower leaders, and, in turn, leaders work to empower their fellow residents. Through these power relations, certain types of active and independent subjectivities are produced by both NGOs and leaders. All of these subjects fit well with the ideology of self-reliance and individual responsibility at the heart of neoliberal governance.

What is empowerment? Empowerment is commonly associated with feelings of personal growth, increased self-awareness, greater respect for oneself, more clearly articulated ideas about what one deserves, or any combinations of the above. It is also often assumed that, once the leaders of a community become empowered, they share this empowerment with the community as a whole, either through example or by sharing skills. My point here is not to reject claims of empowerment through community activism but to follow Raymond Bryant's "general cautionary note to uncritical accounts of grassroots empowerment" (2002, 286). Situations are never this simple, and it is misleading and dangerous to assume that all grassroots- and NGO-sponsored activism leads to empowerment. "Empowerment is apparently bought at a price" (287), and the political outcomes of activism and NGO interventions must be critically examined for their unintended "price" and consequences. My research on colonias points clearly to positive changes in women's lives due to their leadership. It also reveals some much less positive results of the same activism. We do no one a favor by assuming activism always leads to empowerment or that empowerment always leads to progressive social change.

Just like many other widely used and broadly popular theoretical concepts, such as "postmodernism" or "globalization," "empowerment" has become a buzzword that lacks an agreed-upon definition and is often deployed in an uncritical manner. In the more nuanced works I encountered, empowerment was usually defined in relation to the process of gaining life experience that leads to increased personal vision and, in turn, a greater ability to voice one's needs and desires, personal transformation (often political in nature), increased self-esteem, and, finally, more personal autonomy. Jo Rowlands discusses empowerment in these terms: "Empowerment is thus more than participation in decision-making; *it must also include the processes that lead people to perceive themselves as able and entitled to make decisions*" (1997, 14, emphasis in the original). For Rowlands, personal autonomy is clearly related to the ability to make decisions for oneself.

I define empowerment as the process through which people gain a better understanding of their position in power relations, be they domestic, global, local, political, or economic—possibly all of the above. There is one word that is of particular importance in this last sentence and it is the word "process." Empowerment is first and foremost a process, *not* a product, and, for interventions on the part of NGOs to result in empowerment in the colonias or other marginalized communities, this distinction must be acknowledged. In theory, then, this understanding of the processes of one's own marginalization leads to the development of an awareness of how power relations function and how to work within these power relations to affect the change necessary to achieve goals. Like Cruikshank, I believe empowerment is essentially a power relationship in and of itself. As I define it, empowerment closely relates to the processes of *politicization*, or the growing awareness of how political power relations position people based on characteristics such as class, race, gender, and sexuality.[3] Politics are power relations, and for empowerment to have concrete repercussions, it must be political as well. Be it the politics of gender relations in the home or the politics of global development funding, by this definition empowerment must at least address, if not foster attempts at, political change at the individual or collective scale.

In the colonias, the COG often acts as the intermediary between the communities and the state or as a conduit for "the power of the regime." The COG's mission statement reads, in part, "The mission of the [COG] is to improve the quality of life in the colonias-designated communities of southern New Mexico. [COG] works under the following assumptions: *First*, that people are acutely aware of their needs and resources; *Second*, that given the opportunity, people can take positive actions to change conditions in their lives; And *Third*, that by building communities through mutual support, people can become *empowered* to overcome obstacles in their environment which before may have appeared insurmountable" (COG 1998, emphasis mine). If we take the COG mission statement to be a form of empowerment literature, then it falls in the third category of empowerment literature, as it views empowerment as a means to social change and increasing justice.

The COG, like many NGOs, uses the language of empowerment in its mission statement. The idea of empowerment became a popular goal of many development projects in the 1980s and 1990s and "empowerment," is a key term in both international- and domestic-development literatures. Empowerment was the primary goal of most NGO-led development projects that based their political, economic, or cultural agendas on a strong foundation of community empowerment, while empowerment language formed the cornerstone of development practice. Recently, though, this common emphasis on empowerment has been challenged. Increasingly, scholarly work on development criticizes empowerment language, calling into question the very concept itself (Rowlands 1997; Staudt, Rai, and Parpart 2001; Triantafillou and Nielsen 2001). These critiques are based on what is seen as a lack of critical responses to "empowerment" projects and theories that view empowerment rhetoric as a form of governance.

During my fieldwork, I heard both "empowerment" and "politicization" used to describe the goals of the local NGOs. Although the COG uses "empowerment" in its mission statement, the statement also discusses the importance of political participation in reaching the goal of empowerment: "The first area of the [COG's] vision describes an improved quality of life for colonia residents including integration, education and health. Integration into a larger, mainstream society involves participation in activities and organizations that are not isolated to the colonia communities. In this vision colonia residents participate in regional politics, vote at the municipal, county, state and federal levels, participate in non-colonia activities and mainstream civic organizations, and also hold these civic groups accountable to the colonia community organizations" (COG 1998). This statement clearly articulates the process through which empowerment, mentioned earlier in the mission statement, can be achieved. Political participation forms the link between individual empowerment that community organizing can create and social change. Political activism provides the method through which the COG envisions colonia residents holding the state "accountable." Yet as this passage also makes clear, these "civic groups," including other local and federal state agencies, are to be held accountable to the colonias as a collective or to "the colonia's community organizations" and not to the individual residents. This link between political activism and empowerment is crucial for the development of politics in the colonias. Empowerment and politicization can become part of the same processes. In most NGO rhetoric, empowerment focuses on the individual changes in activists' lives, while politicization focuses on social change at a larger scale, such as the community scale we see in the COG's mission statement above. When the two are conflated, as they often are, and empowerment is used as the umbrella term, the distinctions between individual empowerment and community-wide politicization blur. Individuals are empowered to vote as individuals and politicized to challenge and hold accountable civic organizations as a community. Yet, the movement from the individual to community organization rarely takes place in colonias.

Based on my observations, I believe that the COG's main goal is politicization. Much like empowerment, politicization must be seen as a process, that is, the process through which an active political engagement is developed. In community organizing, some NGOs, COG included, base their visions of politicization on Paulo Freire's concept of *conscientização* as presented in his book *Pedagogy of the Oppressed* (2000). In Freire's model, learners individually begin to override their internalized prejudices in order to challenge their daily exposure to the oppressor's hegemony: "Conscientization," as it is known in English, is useful in community organizing because it implies a process, or movement, to a critical awareness and knowledge of one's own oppression and marginalization. This is also often the process of community organizing. In Freire's words, "Humankind *emerge* from their *submersion* and acquire the ability to *intervene* in reality, and it is unveiled" (109). This is why the COG uses Freire as the basis of its organizing.

The term "politicization," used in staff meetings by staff members, best described the NGO's goals. The first step in the politicization process is the individual empowerment of leaders. At this level, the COG saw good results. Yet, by articulating its mission goals in *terms* of empowerment, the organization's other stated goals and daily practices of politicization were obscured. The implicit stress on individual development and resistance that the term "empowerment" carries focuses the attention of both NGOs and community leaders on individual and personal goals. This stress on the individual draws attention away from larger-scale politicization and progressive social change that might make the connections between marginalized groups clearer. The slippage created through the interchangeable use of the terms "empowerment" and "politicization" hides and confuses the lack of politicization at the community scale even more. I argue that until "empowerment" is politicized at the individual scale, NGOs like the COG will find it very difficult to create politicization at the community scale. This chapter illustrates the ways that community projects and NGO interventions often fall short of politicization. The point here is to focus on and document the multiple problems that exist in the relationships between civil society, NGOs, and their client communities. It is, of course, also important to bring into focus the power inequalities that mark the "empowerment" relationship and the very real and central role this relationship plays in the processes of govermentality. One of the primary arguments I make in this book is that NGOs need to become aware of their role in the processes of governmentality and the production of a particular type of active subject–citizen in the working poor communities.

NGOs like the COG need what Hart describes as "a flexible political strategy that recognizes multiple, interconnected arenas of material and cultural struggle; that presses on contradictions and slippages to open new spaces and understandings; that operates simultaneously on multiple fronts; and that searches for connections and alliances" (2002, 33). NGOs require a better understanding of the complex relations between empowerment and politicization if they are to create both, or, perhaps, when they examine their role in power relationships, they will choose not to "empower" after all. Either way, the goals of activism are not being met.

Due to the many relations among activism, empowerment and politicization, and social change, I want to critically examine the development of politicization in colonia communities. How does activism work? Does activism lead to social and political change that confronts systems of power in the colonias? Does activism lead to politicization? And finally, how do NGOs affect the activism of colonia leaders? Colonia activists learn new skills and gain self-esteem and personal respect, yet it is not clear these new skills and characteristics lead to politicization or "empowerment" at the scale of the community as described by COG in its mission statement and through its daily practices.

One Mother, Many Roles

Motherhood is deployed to encourage and motivate colonia women to become leaders and activists and, in turn, through this process often mediated by NGOs, women's daily situated practices of motherhood become those of leadership and activism. Through this relationship, some colonia women become NGO participants and also experience greater connections to forms of neoliberal governance that NGOs may intentionally or unintentionally pass on to their client communities. We see here three key aspects of identity called on for a select group of colonia women: motherhood, leadership, and membership in an NGO. As with any aspect of identity, these three can be called on together, separately, or in combination. At times they work well together, and in other instances they make competing and contradictory demands on an activist's time.

Activist Mothering

The literature on women's community activism documents a growing interest in "activist mothering" (Naples 1998b). Although the concept has been developed primarily in urban settings, I believe "activist mothering" is a useful way to articulate the work of women leaders in the rural colonias. Activist mothering unites the arguments I presented in chapter 2 on how motherhood and family responsibilities produce women as leaders and the arguments I present here on the consequences of women's leadership in their colonias. These strong connections between motherhood and activism actually limit colonia leaders' politicization because they articulate their activism on such a private and individual scale. The development and promotion of mother-centered activism in the colonias serves, yet again, the neoliberal preference for personal responsibility and individual motivations. The mother–child bond is very powerful and can be called upon in powerful ways such that by mobilizing the role of mother and caretaker, colonia leaders address issues well beyond their common situated daily practices. This power can both enable women's activism as a justification for work done outside the traditional women's spheres of home and community and disable it by stressing the individual nature of the mother–child bond, thereby isolating mothers in their activism.

In *Grassroots Warriors: A Study of Women's Community Activism in the War on Poverty*, Nancy Naples uses the concept of "activist mothering" as a central tool in explaining the political activities of the women she studied. For Naples, the term "activist mothering" comes from the women's own words to describe their motivations for their activism: "I recognized how a broadened definition of mothering was woven in and through their paid and unpaid community work which in turn was infused with political activism" (113). The women in my study describe their activism in similar terms. They do what they do because their children and all the other children in the colonia have unmet needs. For these women, it is often hard to tell where their roles as mothers and caretakers end and where their roles as community leader begin. The women that lead the colonias are motivated to do so, in part, because the roles of mother and colonia leader overlap so well. According to Naples, "Activist mothering not only involves nurturing work for those outside one's kinship group, but also encompasses a broad definition of actual mothering practices. The community workers defined 'good mothering' to comprise all actions, including social activism, that addressed the needs of their children and community—variously defined as their racial-ethnic group, low-income people, or members of a particular neighborhood" (113). In this way, even projects that benefit the community as a whole can be described as acts of mothering. The future of the community also falls within this description of mothering: "As residents of poor communities, many of the women described how the deteriorating conditions as well as the inadequate education and health services that threatened their children's growth and development fostered an ongoing commitment to community work" (114).

Naples's definition of "activist mothering" highlights the needs of children, the needs of the community as a whole, and the future of both as central elements of activism for these women. While the women in my sample were mostly identified as mothers acting out of need, they were less likely to articulate these needs in terms of rights and other obviously political terms, as the women in Naples's study did. I argue that this crucial step, from the discourses of individual need to those of community need as a matter of rights and resources, offers a key element missing from women's leadership in the colonias. This step is more than just the move from individual need to community need; it is also the step that shapes how one links one's situated daily practice to the world that is necessary to create a more critical analysis of daily events. In articulating these links, women leaders often looked for guidance from NGOs, who made the connection between women's identities as mothers to their work as community leaders, but did so through discourses of responsibility and tradition rather than conversations about rights and resources or other more obviously politicized topics.

Mothering and the State

Processes beyond the scale of the community also connected the roles of mother and activist. The very same processes of global economic restructuring that made colonias necessary in the first place also pushed women to organize and improve their communities. The rise of the neoliberal political project, which included the cutbacks of basic social services, lead women activists to take on greater responsibilities for the most basic necessities of social reproduction, such as potable water. The activities of the voluntary sector, or NGOs, provide one way in which women, like those that lead the colonias, found help dealing with the new and greater demands of social reproduction. According to Jennifer Wolch, the rise of NGOs is directly related to the changes global economic restructuring makes in social service provision: "The restructuring of metropolitan America has been characterized as a process of spatial change resulting from the reorganization of production and an evolving division of labor. But a second aspect of restructuring is the redefinition of state responsibilities for population welfare. This redefinition, effectuated via public sector retrenchment and privatization of public services, has dramatically altered the organization of the post-1945 welfare state" (1989, 197). A growing group of NGOs was developed to deal with the new demands on voluntary-sector organizations. Wolch calls this group of NGOs the "shadow state." Although these NGOs are called "nongovernmental," the organizations are often, in reality, closely tied to the state on which they rely for funding. Wolch argues that, since NGO politics so closely relate to partisan politics, there are more and less marginalized NGOs, just as there are more and less marginalized constituencies. In the case of the colonias, both the NGOs and the population of Mexican immigrants they serve are marginalized and constantly in fear of depleting funds. Due to the inconsistency and inadequacy of funding for NGOs working in the colonias, the COG and the other NGOs that work in these communities rely, in large part, on self-help projects like those described in the previous chapter. Yet self-help technologies and projects lead colonia residents to do more of the work themselves. In this way, they pick up the slack when the government no longer provides (or forces developers to provide) services, such as electricity, sewage systems, and potable water.

Cindi Katz describes a similar process at work in her discussion of "vagabond capitalism" and its effects on social reproduction for women and children. According to Katz, "Social reproduction is secured through a shifting constellation of sources and encompassed within the broad categories of the state, the household, capital, and civil society. The balance among these varies historically, geographically, and across class" (2001b, 711). By focusing on the shifting nature of social reproduction and on the multiple players involved in this relationship, Katz makes it possible to view the situation in the colonias as a shift of responsibility from the state and capital to the household and civil society. In the current situation, the leaders in the colonias and the NGOs that serve them work together to fill in where the state and capital, in the form of employers, are no longer securing the basics of social reproduction.

The movement of the state and capital out of the arena of social reproduction in Katz's model closely relates to Wolch's analysis of public-sector retrenchment. Katz first notes the state's past role as a service provider and its long history as such: "The state is involved in other political-economic aspects of social reproduction as well. From state subsidies for electrification, water supplies, and sewage treatment to schools and health care services and the provision of a variety of goods and services associated with the welfare state, the state has long been implicated in social reproduction" (713). She then goes on to point to an important change in "recent trends towards privatization [that] have created sharp distinctions between rich and poor households" (713). Finally, Katz pulls together her analysis of privatization and class, using gender to demonstrate how it is often women, such as those in the colonias, who make up for missing resources daily as they struggle to maintain the social reproduction of their households and family: "In many places, these shifts have had a particularly chilling effect on women, who for the most part continue to fill the gap between the state and market in ensuring their households' reproduction and well-being" (713). The women who lead the colonias fit well into Katz's model, as they provide their own water and wastewater systems and other necessary resources in order to ensure the "well-being" of their families. Colonia leaders responding to shifts in the state's allocation of resources organize their colonias to secure the resources necessary to maintain the day-to-day social reproduction of their communities.

Daily Practice and the Limits to Politicization in the Colonias

Community Projects

Adequate heating is a necessity when it comes to raising a family, especially in the very cold desert winters of southern New Mexico. Colonia residents have few choices when it comes to heating their homes. They can buy propane and heating gas in tanks and hook them up to their heater, or they can use electric space heaters. Both choices are expensive. When residents try to heat an old trailer with gaps between the warped floor and the doors, windows that do not close all the way, and most likely a couple of holes in the uninsulated roof, heating bills can run very high. While I was in New Mexico, both Valle de Vacas and Los Montes decided to focus on access to natural gas as their next project. Because the propane truck drivers do not always respond to calls quickly, families have to go without heat when it is below freezing outside. On several occasions, I visited leaders in the winter, and their houses were so cold inside we could see our breath. The leaders always apologized and explained that they had called the gas man, but he said he was too busy to get to them for a couple of days. But, as with all community organizing, the gas project took months to get off the ground, and it was not until one of my follow-up visits over a year later that the first organizational meeting on the gas project took place.

The COG set up a meeting with the El Paso Gas Company (EPGC) so the colonia residents could get an idea of what would be necessary to run lines into their colonias. The leaders went door-to-door asking their communities to come out to the meeting, to prepare questions, and to show their support for the project. As the day of the meeting neared, flyers were sent out with the children after school and the leaders spent more time getting the word out. The night of the meeting, the EPGC representative was there early and chatted with Ernie, the COG representative that evening. The COG's director, Elena, also attended.

The turnout for the meeting was great: twenty-three colonia residents, eight of who were men, were in attendance. The EPGC man, who was fluent in Spanish, spoke first and explained how the gas lines would be laid and outlined the responsibility of each resident. EPGC put together a fair deal for the colonias; if each resident put in their own lines to the street, EPGC would lay the street lines at no cost to the colonia. The residents would only have to pay forty dollars for the meter and leave a ninety-dollar deposit. This sounded like an amazing deal, as in the winter some paid up to ninety dollars every three weeks for propane. But, there was a catch. More than half of each colonia had to sign onto the project for it to be worthwhile for EPGC. When the EPGC representative mentioned the numbers necessary for the project to go ahead, several residents in the meeting had questions. The first to speak was a man from Valle de Vacas who raised questions about the issue of taxes on his lot. After he spoke out, other residents raised their hands and asked if what he said was true: would their taxes go up? The COG and EPGC representatives very cautiously explained that, yes, their taxes might go up but only because their land would be worth more with natural gas available and that was a good thing. But this logic did not relieve the anxiety of the residents, who feared they could not afford the higher taxes.

After the meeting, I spoke to Flora, Marie, and Elena. All three agreed natural gas was a priority and that the savings on gas bills would cancel out the higher taxes. But Marie pointed out that the men were not convinced, and, without their support, the project would never get under way because they were needed to dig the trenches and lay the pipe from the houses to the street. Flora and Elena were both more positive and believed that, with enough information, the men could be sold on the project. I remained quiet, but I agreed with Marie. The men at the meeting had enough second thoughts about the project to keep it from ever getting off the ground. Flora, Marie, and their fellow leaders had organized the colonias enough to get people out to the informational meeting but not enough to get the men to agree to the project. When asked why the men were so negative about the project, the women revisited reasons they had given for earlier failed projects. The men felt they could do more of the work themselves in order to save more money, and, at the same time, they complained about not having enough time to do the work the project required. Of course, the men were also afraid that if the value of their land went up, they would no longer be able to afford it. Elena and I left the meeting feeling defeated and frustrated.

Later that evening we talked more about the situation. Elena mentioned another colonia the COG represented that declined a wastewater system and paved roads because the improvements would have raised their taxes, and the residents believed the colonia was livable as it was. When I asked why they would turn away projects that could not only make their life easier but also protect the environment in which they lived, Elena explained it was about more than just property value and livable conditions; it was about pride. She said she had heard colonia residents say over and over that they did not want the government (including NGOs supported by federal funds) to make the improvements because they did not want to be dependent. My first thought was they are not being dependent but justifiably demanding. The government allowed colonias to develop without proper planning, and colonia residents should not have to pay for the government's failure to enforce subdivision laws. But the residents in the colonias did not think about the situation in this way. At times, colonia residents might step up and make their anger and need known to county and state officials, but then, just as it all seems to come together, they step back and resume business as usual. As Elena said, "there are gusts of involvement" in the colonias. The leaders get very involved and vocal, and then they pull back into the community and their familiar daily practices. According to Elena, that was just the way it went in the colonias. Several months later, the leaders in Los Montes and Valle de Vacas were still trying to get enough people signed on and digging their own trenches for EPGC to begin laying their lines.

These "gusts of involvement" present a serious challenge to politicization in the colonias. The movement of leaders between interest in and detachment from the colonia's projects severely limits activism. The connections between leaders' personal lives and their activism can make leadership a very personal and individual experience. The COG teaches potential activists and leaders that they need individual and personal motivation to be a leader and that this personal motivation is not a bad or selfish thing. Rather, personal motivation is what a leader needs to keep going. Obviously, the leaders I met remained involved because of their specific individual motivations. But the stress on the individual nature of activism, when combined with the COG's focus on material change alone as opposed to changes in ideology, which might link activism to a rights discourse, for instance, makes activism in the colonias a rather isolated affair contained not only within each individual colonia but also within each leader's home and daily situated practices.

The first "gust" of involvement in Recuerdos was a project clearly developed and implemented to help the colonia's children. The children of Recuerdos needed a better street so the school bus could come into the colonia and pick them up. The colonia as a whole needed a better street. The pot-holed and deeply rutted road caused flat tires and damage to cars regularly. But when the leaders remembered the battle to get the street paved, they remembered it as a fight to help the children, not themselves. According to Juana, "The children used to go to school; walking from here. When it was very cold, they would go walking. Even in storms they would go. Then, later, we had a meeting at the school where we started to look for help to find out how we could get a bus to come to Recuerdos. [And then] they started to fix the main road."[4] Access for the school bus is a common first project for a colonia. It is a necessity that appeals to women leaders who are often motivated by the needs of their children, and it is clearly necessary and justifiable to the community as a whole. This project clearly tied community activism to mothering as well and, therefore, was an obvious choice for NGOs involved in partnering programs.

Before it was improved, the road came off the local highway at a ninety-degree angle, which made it very dangerous to turn when the speed at that point averaged more than forty-five miles per hour. Once one got onto the road that led to the colonia, it was necessary to navigate the badly pot-holed road for a quarter of a mile before one arrived at the first major obstacle, the railroad line. There was no crossing light, bell, or bar that went down when a train was on the way. There was nothing more than a small sign that said, "Watch for trains." At the railway crossing, the road barely rose to the level of the tracks, providing for a very bumpy ride over the tracks. Before the road was fixed for the bus, one had to cross the tracks at quite a good speed to get over them, and speeding through a railway crossing is not a good idea. Both Juana and Estella emphasized the battle to get the school bus into the colonia as one of their first fights and one they were motivated to take on because it addressed the needs of their children.

Estella remembered the paving of the road into the colonia as a triumph for the children. She also held up this event as their first real success in community organizing.

> Then in '95 we started to get organized, and, once we had an organization and meetings, after one of our first meetings, COG started to come. Then we asked them to confirm that we had troubles, troubles with the bus. Then they asked us if we wanted to go to Santa Fe to see the governor. And then we said, "Yes." I had my child, who was more or less six years old, and he used to walk on the side of the highway, which is more or less a mile, a mile and a quarter, to get there, and when it rained it was, of course, a problem. Then several people from here got together, and we went to Santa Fe.[5]

Estella was one of the few colonia leaders I met that I can confidently describe as politicized. She understood how her personal empowerment, through new skills and experiences, could be used to create social change in her colonia and to politicize her community. Whenever she was given the chance, Estella challenged the marginalization of her community. She regularly asked why they should have to live without basic resources just because they were Mexicans living in an immigrant community. Yet, she was also often frustrated by the lack of action on the part of her community as a whole. Even with her clear commitment to her work in the colonia, Estella had a hard time involving more people in community matters and, like Flora, Esperanza, and Marie, she found her neighbors disinterested in the politics of their marginalized positions as Mexican immigrant field labor.

Community Organizing and Training: How to Become a Leader

After colonia leaders complete their daily chores and responsibilities, they move on to their duties as community activists. As described previously, most of the women in my sample were chosen by the COG to be trained as leaders. They either exhibited leadership qualities or were in leadership positions before the COG began organizing in their colonia. Most of them received training from the COG on leadership and community organizing, and those who had not been formally trained were indoctrinated with the COG philosophy through meetings and one-on-one interviews with the COG organizers. The COG trainers made it very clear that being a leader was an intimate project and that leadership can affect one's life in very serious ways, both good and bad.

While I was following Valle de Vacas, an effort was under way to redevelop the community organization in the colonia after months of no activity. The first colonia meeting to develop its new leadership was full of people all eager to make change, but when Ernie and Mario, the COG organizers, asked what people were willing to do to organize their colonia, the crowd fell silent. By the third meeting, the COG recognized a core group of three women. These women, Marie, who had remained active over many years, and Alicia and Rosa, both new, became the leaders of Valle de Vacas during the year I was in New Mexico. All three were present in the first informational meeting, but, in addition to Marie, only Alicia and Rosa had shown continuing interest and commitment. We met in Rosa's house; as a single mother this made it easier for her to attend the meetings.

These women went through intensive training with the COG, and I went through the training with them. The four-part training started with a discussion of leadership and the key qualities of a good leader. The group decided that, in order to be a good leader, the ability to listen to others and be sure of one's own words was a necessity. The women associated the qualities of high self-esteem, confidence, and belief in one's qualifications with being good leaders. All the women were actively engaged while the group made suggestions for the list of leadership qualities. But when Mario made it clear that these women had all the listed qualities, which was why they were there, the women became quieter and less interactive. It appeared they had less faith in their own ability to live up to the demands of leadership than Mario had in them.

By the second meeting, the women again seemed very enthusiastic and ready to engage in the training. This session was about motivation in communities. Mario told the women that a leader needs a passion that is all her own; leaders cannot be in it simply for altruistic reasons. The focus on motivations that the COG employs in its training process leads to a necessary level of personal reflection and honesty. When the women they train look into their personal motivations for becoming leaders and how they might or might not benefit from their roles as such, they discuss many personal issues. The bulk of this session involved the women reading their answers to a questionnaire Mario and Ernie gave out at the end of the previous session.

The questions ranged from, "What do you do with your free time?" to "What are your hopes for the future of your family?" and "Where do you want to be in your life ten years from now?" Both Mario and Ernie also answered the questions, and I was asked to do the same. The women's answers were honest and thoughtful, and, by the end of the session, we all knew a lot more about one another. In their evaluation of the training session, the women all commented on how much they learned and how they were going to try to apply these new skills. However, Rosa appeared worried. She feared that with the limitations her disabled son, Carlos, put on her lifestyle, she would find it nearly impossible to get involved seriously. When Mario suggested that her neighbors should come to her house so Rosa could be with Carlos and do her organizing work, Rosa said that would not work because her husband was very suspicious and would ask her why all those people were visiting her at home. At this point, Mario admitted that he saw something in Rosa that made him think she had experienced difficulties. Rosa told him he was right; her life had been hard. At this point, the conversation opened up and Marie made the point to Rosa that "*el hijo es de dos*" (the son belongs to both she and her husband). Rosa acknowledged that this ought to be the case but that her husband does not take any responsibility for the child and believed that the money he sent from his job in California was enough. Marie was not satisfied with this answer and pushed Rosa to acknowledge that her husband ought to be more supportive. Though she never acknowledged that her husband was not supportive enough, Rosa admitted she felt like he neither cared about her situation nor realized how hard it was for her to care for their son by herself.

Rosa went on to say that she had thought about leaving her husband but that she was fearful he might drink himself to death if she did. I would later find out from Rosa that her husband's drinking was one of the main reasons she did not encourage him to visit more. But at the time of this meeting, Rosa had not yet told any of us that her husband's visits were so troubling, and Marie insisted that Rosa make her husband visit more and do more to help her. For many of the women, these sessions offered them a rare space in which they could articulate to non–family members their personal fears and anxieties. These new opinions about their lives could be very informative and motivational. The open and comfortable environment the COG encouraged lent itself to confessions and a therapeutic feeling as the women interpreted each other's actions and anxieties.

Next, Alicia joined the conversation. She, like Mario, Ernie, and I, had been quiet, just listening and watching the conversation unfold. Alicia, with her no-nonsense approach, looked right at Rosa and told her that she liked her situation the way it was because it was "her conscience that kept her in it." Alicia then explained that it was a cultural thing; she believed that women feel they have to take care of the children and that their role as a mother is all that matters. Rosa replied that her oldest child understood the situation, and he was always telling her to leave their father. But a teary-eyed Rosa said she just could not do it. Encouragingly, Alicia said that, if she had to, she would leave her husband and be a single mother to her two daughters; it would be worth it to be happy. Never long out of the conversation, Marie recalled that she consciously tried to break the model her mother had set for her of always bending to the authority of her father and that she has always done what she wanted to do, even when her husband did not agree. In an effort to demonstrate a woman's power in the home, Marie quoted an old Mexican *dicho*, or saying: "*El hombre hace lo que la mujer quiere*" (the man does what the woman wants). I was not at all surprised when Marie used this dicho; I had heard it many times before and had learned that it represented what many women thought about power relations in the home. While men appeared to be in control and assume their patriarchal power with gusto, it was really the women who got their way. It was the women who controlled many colonia households in subtle and nuanced ways, as they controlled their husbands and made them think they got their way while they were actually doing what their wives wanted. After a long round of laughing, all of us commented on how much closer we felt and how we all shared many of the same problems. This knowledge made us all feel better.

Those of us present took part in what could be called a consciousness-raising process. By sharing personal experiences and insights, we began to understand how the feminist statement "the personal is political" is true. Yet, in these workshops and discussion groups we never got past the beginning of this process and to the point where we might help each other to employ these empowering or politicizing ideas. Without any follow-through, the empowerment was not reaffirmed. There was no process in place to check in with Rosa to see if she was able to stand up to her husband nor was the empowerment politicized so the women could see how Rosa's problems with her husband linked to larger-scale issues of gendered power relations in immigrant households. The second session ended with a brief discussion of one-on-one interview techniques and their use in soliciting new leaders and involving colonia residents in the community. Mario explained that by asking questions like those on the questionnaire, the women would understand how a resident might be able to help her community. The women could use the answers they received to tailor a project to a particular neighbor. By the end of the meeting, the women realized how personal community activism needs to be for it to really make a difference. The third meeting focused on what the colonia's first project should be and how the leadership should go about choosing it.

It was in these meetings that I saw a greater potential for politicization. If these women continued to question their situations, practices, and relationships in this manner, they and the community they led could reach an active understanding of their own marginalization and the ways it could be challenged. But the conversation did not continue in this vein. Once this initial four-part training ended, later COG interventions and workshops focused on the mechanics of daily practice and improvements in the colonias. The connections between these necessary improvements and the power relations that created these unimproved communities in the first place are rarely addressed at the scale of the community.

The Neighborhood Politics of Community Organizing

The complex web of relationships that exist in any community makes community organizing very difficult, especially when everyday politics, rumors, and personalities are involved. The daily disruptions and challenges of community leadership occupy much of the women's free time, so I was not terribly surprised to find the women leaders were not tied directly into the greater politics the COG navigates on their behalf.

Once the COG identified and trained core leaders in Valle de Vacas, its job was to help the colonia identify the project it wanted to undertake first. This took considerable time while the leaders talked to their neighbors and consulted the colonia as a whole. Once the need for a wastewater system was identified as one of the most important priorities of the colonia, the leaders had to wage a battle against the county. The county had developed plans for a regional wastewater system that would not be online for another five years, and it wanted Valle de Vacas and Los Montes to join this system. In consultation with a local water and sanitation district, county planners, and a local housing corporation that wanted to do work in the colonias, the COG came up with a plan to hook Valle de Vacas and Los Montes to the nearby Troy Water and Sanitation District. The negotiations for this project took place in meetings in Las Cruces at the COG offices and in nearby restaurants over coffee and breakfast. The COG did not invite or keep the women up to date who led Valle de Vacas and Los Montes. Their role was to get colonia residents to the meetings and make sure the colonias were going to support the project. The colonia leaders' role was very important because the colonias had to show the county that they were committed to the project and that they did not want to be part of the regional wastewater plant. This plant was going to take years to develop and the colonias could not wait that long. Once the Troy option was solidified, the leaders of Valle de Vacas had to settle a disagreement among themselves; Marie did not like the Troy alternative, while Rosa and Alicia just wanted to get the colonia "flushing."[6] After months of meetings and endless discussion, Marie was convinced that the colonia should go to the local water and sanitation district and not wait for the regional plant, even though she personally did not trust the district director. Disagreements among leaders are a daily part of colonia activism and one of the many obstacles confronting colonia leaders. Competing visions for improvements to communities are often the cause of disagreements

such as this one in Valle de Vacas. While in this case a resolution was relatively easy to bring about, as the conflict involved only two competing
visions, other examples of competing visions in colonia leadership can be
more difficult to overcome. In these cases, projects have to be tabled until
leaders can come to an acceptable solution or, in the worst cases, until a
leader leaves her position.

All community activism is political. It is the scale at which these politics take place that varies. The women who led the colonias in Doña Ana
County deal with politics on the community scale, yet they do so as individuals, while politics at the county or state scale most often fell to the COG.
The COG negotiated, with varying levels of involvement from the women
leaders, issues that moved between these two scales. While the women
handled the daily aspects of the improvement projects, they spent little
time discussing the why or how of their situation and the politics that surrounds their communities at the regional and federal scales. They were
often too busy to question the need for these projects and the political
marginalization that led to their lack of basic resources and limited rights.
The women had enough to do, with their responsibilities in their homes
and their organizing activities. The disagreement between Marie and Alicia over the wastewater project in Valle de Vacas was not an isolated event;
leaders often disagreed, and colonias could become split between factions
made up of leaders and their supporters. The politics of community con-
sumed much of a leader's time and patience.

Between two of my visits to Los Montes in May 2001 and June 2002,
the leadership of the colonia changed dramatically. When I left New Mexico in late 2000, Flora and Esperanza were inseparable best friends. When
I visited again in May 2001, I did not see Esperanza, and I was surprised
to see that she did not stop by Flora's while I was there. Her name rarely
even came up during the visit. When I asked Flora about the change in her
relationship with Esperanza, she said they did not talk any more. Esperanza "was on Lola's side." I knew who Lola was. She had been a leader
before I got to Los Montes, and when I had asked the COG about her,
both Ernie and Elena steered me away from her and pointed to Flora and
Esperanza as the current leaders with whom I should talk. I tried to meet
with Lola, but she was not interested in my project, and I finally decided
that she was removed enough from the current leadership that my work
would not be affected by her absence. As it turns out, Lola had disappeared from the daily workings of the leadership in Los Montes but had

not stayed completely out of the loop. She had recently stepped back into organizing. Lola was convinced that the COG and Ernie were not doing enough for Los Montes and that the colonia would be better off without the COG. Flora did not agree. She believed that the COG helped Los Montes in many ways and that Ernie was a good person and a good friend to the colonia. Esperanza, according to Flora, was not sure whom she agreed with at first, but with a little time she went over to Lola's side and was also calling for the COG to do more or get out. Flora was afraid that the community would be split or, even worse, that the COG might believe that all of Los Montes felt this way, and then Ernie would stop working there. Then Los Montes would not be able to continue the improvements they had already begun. While discussing the situation over chile rellenos at a local bar, Flora expressed a great deal of concern that Lola should not be seen as the voice of Los Montes. Flora and many others believed that the COG did good work, and that Los Montes was indebted to the COG. Flora was very worried and lost a lot of sleep over this situation. Politics such as this are common to both community and NGO activism, and colonia leaders must often deal with changes in personnel at the NGO level while also negotiating changes in leadership at the scale of the colonia. By 2007, Esperanza and Flora were friends again and the three of us had a cold drink on a very hot summer afternoon while discussing the colonia's recent activities. Like any other community, colonias are full of gossip, intrigue, and shifting friendships and alliances.

Effects of Activism on Women Who Lead

The above stories illustrate that through their activism, the women who lead the colonias expand and reorder their daily lives to include meetings with NGOs and county and state officials, and to attend to their leadership duties. But what effect do these new activities have on their lives?

Skill Acquisition

When asked if activism changed their lives, nearly all the women in my sample first answered something along the lines of, "No, no, I'm the same." I quickly realized this question did not elicit the results I hoped for and required reworking. After all, I saw changes in these women in the short time I had known them. They became more self-confident and more inquisitive. Surely they could see these changes as well. I soon found out that if I reworded the question to say "What have you learned from your activism?" the women had plenty to say. Though they admitted they learned from their work as leaders, they were not quite as willing to admit their leadership activities changed them as people. I have argued that motivations for activism, and what women get out of their activism, are closely tied to how women are produced as leaders. All the women with whom I worked became more confident in their communication skills during the period I worked with them. For example, they showed greater comfort addressing public officials and groups of NGOs visiting their colonias. Many women also pointed to their new and improved people skills as a benefit of their activism (Feldman, Stall, and Wright 1998).

One of the most common new skills acknowledged by the women was their improved ability to listen to and talk to people. Flora not only acknowledged this improvement but also tied it to her work for the community: "Yes, I've changed, I've changed. My character has stayed the same, yet while working for the community, I've changed in that I've learned to live together with people, and before I did not know this. I didn't know how to talk [to people], and I felt less able to do it."[7] It is interesting to note that although Flora does admit to having "changed," she makes a point of saying her "character has stayed the same." There was a deep-seated need for many of the colonia leaders to protect their character or what they saw as some inherent aspect of themselves that might somehow be jeopardized or negated by activism.

In much the same way that Flora learned how to talk to and live with people more easily, Juana also describes learning to live with others: "Look, you have to teach people to get along with everyone, angry people and all, to listen to people and also to like to be listened to. Because if you are going to be in a meeting and you want something, you have to teach people to treat each other well so you will be treated well. That's one of the things we learned from COG—how to lead people."[8] The skills that Juana and Flora learned from their activism helped them as leaders and in their personal lives. They became better able to communicate with their families and their friends, and, as Esperanza explained, they were better able to stand up for themselves: "I was also very shy and didn't defend myself from people, you know? My in-laws insulted me and I never defended myself. I don't cry anymore, right? I'm rebellious. If someone says something, I talk back. I don't let it go."[9] Yet, Esperanza did not attribute the change in her character just to her activism; she also pointed to her separation from her husband as a key event in her development as a person. It is interesting to note that these two events developed together. During her separation she kept herself busy with her community activism.

Juana also described her activism as contributing to her ability to deal with people, and she wanted others to learn the skills she learned: "Because I want other people to learn as well. Because it is great to know how to talk, to know how to say what you need to say, [about] what you don't have."[10] The ability to talk more easily with people and to stand up for oneself is a result of activism. Although these results manifest themselves on an individual level, they can also benefit the community as a whole, especially when the leaders have a desire to share them as Juana did. This could be seen as an example of "empowerment" moving from the scale of the individual leader to that of the community as a whole.

Increased Self-Confidence

Increased self-confidence was a consequence of the achievements, new skills and activism in which these leaders took part (Cable 1992; Naples 1998b). It is also this increased confidence that keeps many women active. Along with results, personal growth and increased confidence are the most powerful motivations to continue activism (Feldman, Stall, and Wright 1998).

Estella was one of the most outspoken of the women in regards to her personal growth. She also attributed personal growth to her experiences as a leader. When asked if she had changed because of her activism, Estella was one of the women who responded, "Yes": "Yes. Before I always liked to talk with people a lot, but now I have more confidence in myself." Estella described at the individual scale what these changes meant for her activism: "Before it was difficult to talk with people that came here and to express myself and ask for the things the community needed, and now that I am more sure of myself. I know what I want and that I can get it." Estella knows her work in her colonia and her ability to make change possible puts her ahead of other women: "Now I have proposed things and gotten them, and, for me, that's a lot because not all women can say that they get things done."[11] Estella acknowledged that she was more capable than many other women, but she also acknowledged that this confidence did not just stem from her activism. It also came from the support of her family and her husband. Estella's husband was very supportive of her activism and this enabled her work. Most of the husbands of the married women in my sample were supportive, although to varying degrees. As one of the most important motivations for Estella's activism, her children and their support were central to her leadership: "Well, to have confidence, I believe you also have to have support. For example, my husband always, well, he laughs at me but says, 'you can do it.' And then my children say to me, 'All right, mama!'" [12]

Personal growth not only led to increased self-confidence but also strengthened a number of new skills, such as patience:

> Have I learned? Yeah, you learn to deal with certain things. I mean it's not comfortable to open the door and have all this just because we have no payments [reference to the low monthly cost of buying a colonia lot] and all these things, or the smell of this and that, all these things around us. You learn to say, "Is it that bad?" No, it's not. Can we still live here comfortably and not worry about all these other things? Yeah. So I've learned to just deal with things. I can't change things at this point, and maybe that's just another thing that I've learned. I want things to get done right away, and I know they're not going to get done right away. So patience [is what I learned].[13]

The resigned acceptance in Alicia's tone is common among colonia leaders. They are active in their colonias and believe they can create change, but they know well that nothing changes quickly. As Alicia says, they learn "to deal" with things the way they are because it is necessary in order to survive. Colonia leaders work within a needs-based discourse most of the time, and Alicia's words demonstrate both that and the sense of bounded agency they face daily.

The Simultaneously Enabling and Disabling Nature of Women's Activism

The "dealing" that Alicia refers to is common in the colonias, where residents "make do" with a continual series of ingenious and sometimes dangerous makeshift alternatives for the basic resources they lack. The leadership of women like Alicia creates change regularly. It brings in the

much-needed infrastructure and resources that make daily life much easier, safer, and healthier. But Alicia's leadership, as well as that of her fellow
leaders, creates less progressive social change that challenges the systems
of power that marginalize colonias than would be expected based on their
level of involvement and dedication. How is it that women such as Alicia are able to be activists and leaders who produce important and necessary infrastructural changes in their communities, yet the production of
meaningful discourses surrounding why these communities do not have
basic services and resources in the first place appears to be disabled at a
very basic level? The answer to this question is firmly entwined in the daily
situated practices of the women and NGOs that work with them to organize the colonias. NGOs identify and train women leaders who, in turn,
organize their colonias to create almost entirely infrastructural change yet
do little to rework beliefs about and causes of the poverty, discrimination,
and marginalization that mark these immigrant communities. This is the
community-level work that NGOs and leaders have trouble creating.

While shared experience may enable women leaders to work together
for improvements, it simultaneously disables their efforts. The very same
experiences of poverty and need that encourage women leaders also
limit the time they have for activism and generally drain their energy
and personal resources, thus limiting their leadership efforts. While
need and lack are powerful tools to use to unite a community, they present extremely difficult obstacles to community unity and organization.
During a recent return visit to Los Montes, Flora explained why, seven
years after my initial research, it was still just her who showed up regularly for community-organizing events. I was pleasantly surprised as I
drove into the colonia along Los Montes Road (see Figure 3.1) to see
the community was in the midst of a full-scale paving project that would
pave all three main roads. Flora explained that she and one other woman
had done all the organizing work to get the county-funded project off
the ground and that they had gone door-to-door and collected a signa-
ture from *every* household in the colonia, yet *no one* else was willing to
help. Flora told me that daily life was hard, and people had nothing left
with which to work on their community at the end of the day: "They
want changes, but they want someone else to do it." Flora continued by
saying that she *wanted* to do the work. The work itself seemed to animate
her, and that was obvious as she talked about the paving project and the
other projects she had been involved in during the time we had known

Figure 3.1 The new road in Los Montes. Without the hard work of a few select women, such improvements would not be possible.

each other. How she got to this point was not as clear to her; all she could offer was that she always knew she wanted to work for her community and that this was part of her calling.

NGOs like the COG both enable and disable women's leadership as they focus their interventions in the colonias on the responsibilities of women as mothers to care for their families and community. The data presented here demonstrate the many ways in which the COG enabled women to be activists and leaders who created important changes and improvements in their colonias. Through leadership training and community projects, the COG helped women gain new skills and self-confidence, enabling them to be more active and take on a variety of projects. Yet the collaboration between the COG and the colonias appears to have led only to physical improvements and to little long-term, progressive social change that confronts the marginalization of these communities. The NGOs that work in the colonias both enable and disable the leaders with whom they work in their ability to politicize their communities, and, in turn, these leaders can only create limited change. This contradiction in NGO practices and results directly relates to the position of NGOs within the neoliberal political project. As we have seen in their roles as intermediaries with the partnering state and in self-help programs, NGOs often serve as conduits for neoliberal governance.

To summarize the connections that lead to a lack of progressive social change in the colonias, we start with the striking interconnections between the enabling and disabling factors in the daily life practices in the colonias. The poverty and NGO interventions that so clearly motivate women to become leaders are also the two key elements that disable activism and social change in the colonias. The widespread poverty and need that are so common in the colonias make life difficult so that seemingly simple tasks can become overwhelming and discouraging. People who live in poverty are tired, fearful of unnecessary risk, and very focused on their day-to-day needs. Yet the sheer weight of these tasks can, and does, motivate some women to work for change. NGOs identify these motivated women and use them to organize the colonias in order to create change. But in most cases, this change is almost entirely infrastructural and does little to reshape beliefs surrounding, or understandings of, poverty, discrimination, political and economic marginalization, and their causes. To create a more politically enhanced environment in the colonias, it would be necessary to unite the processes of politicization with the goal of meeting daily needs so that material and social change can be produced together and affirm each other.

To Empower or Not to Empower

"We may not make revolutionaries."[14] That is what Elena said about the work the COG does at a staff meeting I attended in the summer of 2005. I was back in Doña Ana County to discuss an article I had recently finalized for a special issue of a geography journal on the professionalization of NGOs and wanted their feedback on the piece. Although it had not been COG staffers who had first opened my eyes to the lack of politicization in the colonias,[15] I had spoken to Elena and other local colonia activists and NGOers about my concerns over the last few years, and they had worked with me on my analysis and theorization of the situation.

The question being discussed at the staff meeting was whether the COG and other NGOs "disable" activism and, if so, how? Marlene, the grant writer, was the first to say she thought this idea might put too much weight on the role of NGOs in the process of community organizing. It was true, she commented, that NGOs such as the COG did produce leaders, but she questioned whether their actions after that really went on to disable activism. Marlene is a very well-read and educated woman, a writer who teaches creative writing when not working on grants for the COG, and her grasp of the neoliberal politics and economics at play in the colonias was one of the strongest in the group. She went on to suggest that it might be that the NGOs were not as instrumental in keeping people from activism as neoliberal rhetoric was in keeping them tied to self-help. Marlene lamented that "it's hard to generate activism at a community level," then questioned why "the neoliberal rhetoric is so effective at every level." In response to her own question, she responded, "They [colonia residents] believe the bootstraps thing should work."[16] Drawing on the American Dream, that anyone can make it as long as they work hard enough, Marlene made her point that it is neoliberal ideology, and perhaps not the NGOs, that is disabling activism.

It was at this point that Elena said that the COG might not be making revolutionaries, which may not sound like a shocking revelation. It may even sound fairly realistic for a small community-organizing NGO. Yet, nonetheless, the COG staff hopes to create change on this scale daily. After Elena made this statement, Marlene softly said, "We haven't totally accepted this yet," and the room grew quiet for a few moments. And then, as if the floodgates and been opened, the staff began to discuss the idea of politicization and whether it existed in their partner communities. One staff member agreed with Marlene and Elena and volunteered that she did not see the women as very politicized and certainly not the communities. Others pointed to the long-term nature of community organizing and thought the COG was doing OK for the length of time they had been working in these communities. Elena concluded the discussion with her ideas on the root of the problem: follow-up. Because of the nature of activism in the colonias, characterized by the gusts of involvement, the COG had quite a bit of trouble creating what she described as a "social–political space" to foster politicization.

I thought it was very interesting that the COG also saw a missing element of politicization within the colonias but that they did not appear to see any connection between themselves and the rhetoric of neoliberalism that they found so disabling. They would openly discuss the limits of accountability and professionalism on their daily operations and situated practices in one breath, and the limits self-help puts on political awareness in the colonias in the next, but not connect the two through their own efforts. It is hard to say if the missing step was due to a lack of self-reflection or a desire to simply not acknowledge this reality, but, either way, movement beyond this impasse was not happening. I think what was, and still may be, required is for more reflection to occur on the dynamics that are thwarting the production of the revolutionaries they would like to be creating.

Conclusion

The clearly limited results of women's activism and leadership presented a very discouraging discovery for me. During my year in the colonias, I felt surrounded by powerful women making amazing strides to improve their marginalized communities. It was not until after I left New Mexico and was able to look back with an eye for the political results of activism that the full, less-than-encouraging picture became clear. In the hectic planning meetings prior to a project's implementation, the leaders and NGOs rarely had time to cover all the necessary details, and there was rarely time left for discussions of the big-picture politics of the different projects and how these might be made more salient to the community as a whole.

Women's daily experience of activism and leadership rarely led to politicization on the community scale. Some women became noticeably empowered as individuals and reworked transnational gender relations in their homes to reflect what they believed to be more egalitarian situations. But their activism did not lead their colonias to a greater understanding of how to comprehend and challenge their own marginalization as part of a larger population of Mexican immigrants. This lack of politicization directly links to the simultaneously enabling and disabling nature of activism in the colonias. Important elements of women's activism, such as the great motivation poverty precipitates and the influence of NGOs, work to both encourage and produce women predisposed to leadership and to limit and suppress the work of these very same leaders.

The simultaneously enabling and disabling discourses of activism resemble the discourses that produce colonia leaders as subjects who serve the interests of neoliberalism. When the state and NGOs produce colonias as a logical and practical solution to the lack of affordable housing for Mexican working poor, the production of leaders as capable and ingenuous people who can provide for themselves fits perfectly into this neoliberal model. I have argued that these women are capable and resourceful, but they are also overworked and overextended. These latter two qualities, along with the neoliberal ideology at the center of colonia development, mean leaders have little extra time or energy to devote to doing political work, such as learning more about the marginalized position of their own colonias and politicizing their communities. When the fastest way to get drinking water is for the community to lay the lines themselves, colonia leaders will organize their communities to get it done. This is especially true when the discourses that produce activism and activists dictate self-help as the best method, even if this means essentially buying into the neoliberal system of personal responsibility and accepting in turn the services to which your community loses access as a consequence.

The many projects that involve colonia leaders include building wastewater systems, paving roads, gaining access to clean and potable water, installing natural gas, creating community parks, and organizing community-cleanup days. Clearly, the women who lead the colonias create tangible results. But what of the cost? Many projects cost colonia residents little more than their free time and labor, but, for the working poor, free time is precious, and physical labor power is what they rely on for work. Expending their energy on their day off can leave them too tired to function properly the rest of the week.

This chapter addressed the questions of those affected by the activism of these women leaders and the kinds of change that those affected experience. Significant progressive social change that confronts systems of power is limited, most of the time, to the leaders and does not affect the colonia as a whole. Therefore, the kinds of change that colonia leaders experience is the most important question to consider. As I have shown, the leaders acknowledge gaining new skills and greater self-confidence, but few directly articulate gaining a better understanding of how their oppression as colonia residents functions. The leaders may have a stronger idea about the causes of their marginalization, but, without the ability to articulate this idea, the empowerment process is, as Rowlands describes it, only partially complete (1997). If leaders are not expressing an increased personal sense of political awareness, then it would be hard for them to politicize their colonias.

On a return trip to New Mexico, I went out for beers with Ernie and Sarah. I had just recently begun to look at the consensual aspects of colonia activism and the contradictions of women's leadership. Ernie and Sarah are both college-educated activists with long-term commitments to community organizing. As with many activists that work in the field and not in the academy, they often express contempt for academics and social theory. They only accepted me, so they said, because I worked side by side with them for a year, which distinguished me from most of the academics they worked with in the past. But even though they liked me and we got along well, they had no time for the theory I employed in my work. As frustrating as it could be trying to convince them of the value of social theory, it was also always a great way to work through my theoretical ideas. On this particular night at the local microbrewery, Ernie's skepticism reaffirmed an idea I had been mulling over.

We were discussing a meeting we had attended the night before at which a woman asked why the county did not provide natural gas to her colonia when it provided it to everyone else's neighborhood. Ernie used this example to show that colonia residents sometimes do understand the way the system works, and they can see they are being ignored and they do resent it. I agreed with Ernie and then asked why he thought it was that the residents did not follow through with their complaints and try to directly change the politics of their marginalization. Ernie's answer made me realize that I had known the answer to this question all along: "These people did not come here to fuck shit up. They came here to work and have a life."[17] He continued to say that it was not fair to ask the poorest people to change the system. I had often thought about how amazing it was that a single mother like Flora could actually find time to be a leader as well. There simply are not enough hours in the day to add much more to her list of responsibilities, and this is the way many of the women who lead the colonias live. They already had too much to do to make up for the daily racial and environmental discrimination they experience.

It was interesting that Ernie took this position because he and Elena shared an interest in Freire's transformative pedagogy. This discrepancy between Freire's theories of change from the bottom up and Ernie's experience with overburdened colonia residents is very important and leads to the themes discussed next. Theoretically, the NGO bases itself in the teachings of Freire and ideals of empowerment and politicization for the poor and marginalized. Yet, as I have discussed, NGOs like the COG find it difficult to operationalize these interests and desires.

Why is it the case that colonia activists are being politicized on an individual scale and still find it hard to spread this politicization to the community? The way these women are introduced to and trained in activism shapes the practical way in which they approach community activism and the creation of immediately necessary material projects. The contradiction between the theories of politicization, in which the COG and other NGOs are grounded, and the COG's daily practices that value and *prioritize* practical and immediate solutions constitutes the central topic of the next chapter.

CHAPTER 4

The Place of NGOs in Daily Life

"¡Conozco mis derechos!"
I know my rights!

"¡No firme nada!"
Don't sign anything!

T HEIR VOICES CARRIED CLEAR across the parking lot and nearly out onto the interstate, which ran close by the new community center in which the crowd, mostly women, was sitting. They were enthusiastically practicing the proper response to questioning by *La Migra*. This was just one small aspect of an all-day immigration-rights workshop. If one were to walk past the building and did not know what was going on inside, one might think there was a mass arrest taking place.

It was late fall, and more than twenty women and two men had turned out to the first in a series of immigration rights workshops the community-organizing group (COG) planned. This one took place in the colonia of Del Cerro beneath a breathtaking mural of the cycle of life in the Rio Grande valley (see Figure 4.1). The mural was full of images that really hit home with colonia residents. As the women sat below images of indigenous and Mexican immigrant field laborers on a backdrop of fertile farm lands irrigated by the river itself, they listened to Elena tell them that the day was not about their papers or documentation status but rather about their civil rights. I wondered, as I took in the myriad details of the mural, if the irony of the situation was occupying the thoughts of the others in the room as well. Here we were sitting below a mural that clearly celebrated the contributions of Mexican immigrant and indigenous farm labor while we struggled with how to avoid and document abuses to the civil rights of the same groups. The image was filled with the fruit of this labor on multiple levels, including the crops it depicted in the fields and the rainbow of local crops floating across the sky made up of chile, both green and red, onion, alfalfa, cotton, and pecans. Finally, there was the woman lying in

Figure 4.1. Mural in Del Cerro Community Center, where many local colonia community-organizing meetings take place. Del Cerro is a colonia housing many immigrant farm workers like those depicted in the mural.

the field giving birth to a child. It is true; the fields give work and life to farm-worker communities and colonias in the valley.

The need for the training that day came from a long string of reported and documented abuses by *La Migra*. These workshops were an effort to help colonia residents learn how to both defend themselves by using their civil rights and to teach them how to document inappropriate and illegal searches on the part of the border patrol. As solely Mexican communities, colonias were often targeted for searches by the border patrol, and, without a basic knowledge of their civil rights, residents were much more susceptible to harassment.

In this chapter, I tie together, through a discussion of the social production of colonias as racialized spaces, previous arguments about the social production of the Mexican working poor on the border, the limits of women's activism, and the development of colonias. In the process, I add a deeper and more critical account of the role nongovernmental organizations (NGOs) can play in colonia communities as well. Through examples from the work colonia leaders do, the hegemonic production of space through racism and economic and social discrimination comes face-to-face with the counterhegemonic construction of the just and well-developed space for which colonias leaders and NGOs strive. As this chapter traces the relationships between the many forces at work in the colonias and their roles in the social production of colonia space, I ask the question, what is the role of NGOs in the development of progressive social change and resistance in the colonias? In response, I argue that the slow growth of social change and resistance in colonias can be attributed, in part, to the growing partnership between NGOs and civil society and the inadvertent outcomes of NGO interventions. This answer is developed through a concrete examination of colonia and NGO practices and moves from these material daily practices to a discussion of social reproduction as it plays out in daily life.

Through the use of specific practices as examples, this chapter focuses on the scale of daily life in colonia communities. It becomes clear that quotidian practices are constitutive of scale, and the practices of colonia residents and the NGOs that work regularly with residents in the colonias construct the scale of daily life in colonia communities. A concentration on the concrete activities of colonia residents offers insight into both the processes and "language[s] of spatial difference" and the ways in which "the production of geographical scale is the site of potentially intense struggle" (Smith 1993, 62). This is especially true as the processes of the "rescaling of governance" move social service provision from the state to civil society and NGOs, which use women leaders as the entry point into many communities including the colonias (Larner and Craig 2003, 1; Wolch 1989).

Geographies of Isolation and Segregation

Chapter 1 outlined the political and economic reasons for both the development of colonias and their physical isolation. I discussed the manner in which the combination of an inadequate supply of affordable housing, lax subdivision laws, and the inability of Mexicans immigrants, recent and more established, to compete in traditional housing markets led unscrupulous land developers to create colonias. On further examination, it becomes clear that the global flows of people and capital that enabled the creation of colonias are marked by race and that race plays a large role in the creation of colonias and in the discourses that surround these Mexican immigrant communities. Colonias are racialized and impoverished communities. The poverty and resource deprivation that exists in the colonias directly ties to the racial and ethnic discrimination that colonia residents experience as Mexican immigrants. The daily experiences of colonia residents also connect to the growing influence of the neoliberal political project that leaves communities like the colonias without basic services. In this chapter, I discuss how this political project, through its ideology of self-reliance and personal independence, engages with civil society to shape a particularly self-sufficient citizen and how this relationship plays out in the daily interactions between residents and NGOs.

In the colonias of New Mexico, the ties between race and the economy become especially clear when one keeps in mind the factors that led to the development of the colonias. Though Anglos also have problems finding affordable housing, it is not a surprise that colonia developers do not target them. Mexican immigrants remain the focus of colonia development not only because they cannot access the traditional home buyers' market but also because they are Mexican and, as Mexicans, they are discursively constructed by non-Mexicans as needing and expecting less.

In his work on identity formation along the border, Pablo Vila examines the prevailing discourse in the U.S. Southwest that suggests "all poverty is Mexican." According to Vila, "Poverty, for many people, is synonymous with Mexican" (2000, 84). Vila's research with residents on both sides of the border around the El Paso–Juarez area acknowledges that both Mexican and Anglo Americans often believe Mexico harbors the worst kind of poverty and that those parts of the United States along the border that contain similar poverty are, in some way, related to Mexico. When discussing my work in colonias, I often encountered discourses that tied the impoverishment of colonias to Mexican poverty even though the colonias exist in the United States. These discourses naturalize Mexican poverty and leave no room for its analysis or for more structural explanations. Newspaper headlines around the El Paso area routinely describe colonias as "the third world in our backyard" (Hill 2003). Many of the Anglo residents I met agreed with those Vila interviewed in their belief that Mexicans like to live in poverty because they are lazy and cannot be bothered to improve the conditions in which they live (Vila 2000, 111).

The space of the colonias is a Mexican space. Most colonia residents are Mexican, or of Mexican descent, and the impoverished spaces of the colonias themselves are labeled "Mexican" by outsiders. The question becomes, which came first—the "Mexican-ness" of the colonias or the poverty? Chapter 2 demonstrated the complex ways in which race, politics, economics, and "tradition" all worked together in the development of the colonias. Here, I will study more closely the relationships between the production and control of space and race. I agree with Donald Mitchell that it is important to examine "the degree to which 'race' itself is a project of the ordering and controlling of space—and of ordering and controlling the movement (or 'travel') of people" (2000, 258). Colonia spaces are marked as Mexican and therefore considered poor, underdeveloped, and dangerous. At the same time, people also view them as resource deprived, backward, and unfit for Anglo Americans to inhabit. In this way, the two sets of related discourses, that the colonias are Mexican and therefore poor and dirty, and the colonias are poor and dirty and therefore only fit for Mexicans, reinforce each other to create a racialized space, a space in which "Mexicans" can live and with which "Americans" need not be concerned.[1] In a piece on El Paso's colonias, The *El Paso Times* quotes a developer as saying, "Curbs and sidewalks, people out there don't want that stuff" (Hill 2003, 148). According to the *El Paso Herald Post*, "People who go there [colonias] make a conscious decision," thus implying colonia residents choose the poverty of colonia life over more-developed communities (149). The binary this discourse creates racializes Anglos against the poor, dirty, Mexican Other.

But how is the space of the colonias marked as Mexican? Is it just because there is no running water or because there are no paved roads? That cannot be the answer because many rich Anglos in New Mexico live in areas that are also supplied with well water and accessed by dirt roads, and these are not "Mexican" spaces. The most prestigious neighborhoods in Santa Fe, New Mexico, one of the state's wealthiest cities, are those with dirt roads. Colonias are Mexican spaces, in part, because they are filled with Mexican bodies and the body is the most fundamental scale at which spaces are experienced, produced, and constructed, and it is also the primary scale at which race is socially constructed. Local discourses dictate that when brown bodies live without running water and paved roads, they do so because they want to and because they are "Mexican" and are used to living that way. To be racially or culturally Mexican means to be predisposed in some way to live in poverty or to live in a colonia.

Civil Society and NGOs: The Role of Nonprofit Organizations in Producing Colonia Spaces

Capital requires Mexican working poor as labor along the border. The Mexican working poor require a place to live, raise families, and make communities. The neoliberal political project requires someone other than itself to provide the Mexican working poor with basic resources. According to Helen Hintjens, "The state is no longer to be held accountable for ensuring that citizens' basic needs are being met; instead private citizens, individually and collectively, are expected to provide for themselves, however poor or disadvantaged they may be" (1999, 386). Between colonia developers and NGOs, all these needs can be met—some better than others. The state and capital can step back, while the developers and civil society get the job done.

In the global south, NGOs have been doing what many consider to be the work of the state for decades, providing basic social services and infrastructure to working-poor communities. Giles Mohan creates his own working definition of civil society and its four key characteristics to use for NGOs in Ghana. Mohan first theorizes civil society as "constituted across local, national and international territories" (2002, 135); this geographical insight is an important element in his argument because of the international aspects of much NGO development work. Mohan argues, "Civil society does not lie unproblematically between the household and state, but shapes and is shaped by economic forces" (135). He situates civil society with a Marxist frame, focusing attention on the importance of economics and the role of the relations of production. Like Gramsci, Mohan implies the existence of hegemony between the state and economic forces that serve the state: "If we accept that the state and society are not separate spheres we must examine the shifting processes of rule operating in the 'hybrid' interstices between them" (135). Mohan knows many NGOs exist and work in these "interstitial spaces," and he points to the importance of locally situating debates on civil society and its functioning, especially since much of the literature on civil society comes from Western experience. In the case of colonias, I do the opposite and take examples from debates over civil society and NGOs from the global south and apply them to some of the poorest communities in the global north. Clearly, the role of civil society is crucial to understanding the relationships between NGOs and the state.

In his work on the state and civil society, Nikolas Rose points to the increase in discourses surrounding the role of individual citizens in neoliberal theory. He makes the connection between the development of the self-reliant and self-sufficient citizen in neoliberal policy and its attending processes of shrinking state interventions and decreasing welfare. In this new era, "the relation of the state and the people was to take a different form: the former would maintain the infrastructure of law and order; the latter would promote individual and national wellbeing by their responsibility and enterprise" (Rose 1999, 139). The role of the "social state" gives way, according to Rose, to the "enabling state" (142). The "enabling state" no longer takes responsibility for the intricacies of daily life, health, and security. Individual citizens and the "third sector" now assume this role. Rose's "third sector" encompasses much of what others call civil society and the voluntary sector. It is the "third sector," in the form of the COG and other NGOs, that takes over in the colonias.

The neoliberal cutbacks that follow global economic restructuring affect the colonias as they have many other poor populations. These cutbacks hit colonias especially hard because they are accompanied by further reductions in social services as a result of growing nativist sentiments. Though "nongovernmental," the NGOs that provide the services that the state now fails to provide are deeply implicated in the state's campaigns for individual self-reliance and governance. In the case of the colonias in Doña Ana County, the COG does not provide direct services, but it does supply information on service providers and organizes the colonias so that they can provide for themselves.

As the state steps back from its role in welfare and social service provision in order to implement the economic policies of the neoliberal political project, civil society is in a position to take over some of the social service provision. In the neoliberal framework, this is how social service provision is meant to work. Mohan refers to this as "the period of neoliberalism when the NGO sector blossomed" (2002, 139). As part of Wolch's "shadow state," these NGOs take the "form of voluntary organizations with the collective responsibilities, which operate outside traditional democratic controls, yet are strongly affected by state resources and constraints" (Wolch 1989, 198). The sources of funding they solicit and the ways in which the federal government often controls this funding determine, in part, the activities of the shadow state organizations. The COG is one such organization. Although the COG is very careful about where it gets its money and goes out of its way to solicit funds from foundations that share its progressive ideas, it is still part of the shadow state. The relationship between the COG and the state is quite complex. Much of the funds the COG helps procure comes from federal government agencies like the U.S. Department of Housing and Urban Development (HUD). This money often comes with limitations and directives. When I last visited the COG, they had three federal construction grants, all of which came with many stipulations and a mountain of forms and reports.

James Petras and Henry Veltmeyer take a more cynical approach to the role of NGOs in the processes of global economic restructuring. They believe NGOs "mystify and deflect that discontent away from direct attacks on corporate/banking power structures and profits and towards local micro-projects, apolitical 'grass roots' self-exploitation and 'popular education' that avoids class analysis of imperialism and capitalist profit-making" (Petras and Veltmeyer 2001, 128). Petras and Veltmeyer theorize the role of NGOs in civil society as tools of the state, tools that address and provide necessary programs of social service provision, yet do so in the least challenging way possible. By focusing on the immediate needs and daily desires of their clients, NGOs, according to Petras and Veltmeyer, avoid the more political questions of rights and resources that lie at the heart of social service cutbacks. Similarly, Mohan argues, "Donor and NGO support for civil society, and 'localism' in general, keeps at bay debates about more fundamental structural changes to, say, unequal property rights or despotic, but economically useful, host governments" (2002, 150).

In the colonias, leaders expect to see COG organizers at least once a week, and residents do not look twice when COG organizers drive by with potential donors or county officials. During one of many door-to-door campaigns in Los Montes to interest a colonia's residents in a new project, I was struck by how uninterested residents were. Residents are so acclimatized to the workings of NGOs that the presence of a young gringa walking through their all-Mexican community barely provoked interest. Most colonia residents simply accept NGOs as part of their life, and it is often through the workings of NGOs that colonia residents experience neoliberal policy measures (Dolhinow 2005).

To understand how NGOs became so commonplace in the colonias, it is necessary to examine the role civil society plays in the current neoliberal, or "advanced liberal," order (Barry, Osborne, and Rose 1996; Dean

1999; Rose 1999). Two of the most powerful theoretical approaches to the state and civil society are those offered by Antonio Gramsci and Michel Foucault. Although these two theorists address the powers and processes of the state from very different points of view, they can be used productively together to conceive of civil society as simultaneously "both object and end of government" (Burchell 1996, 25).

Together, Gramsci and Foucault's theories of the state and governmentality work well to theorize the often-contradictory role of NGOs in colonias. According to Gramsci, "The State is the entire complex of practical and theoretical activities with which the ruling class not only justifies and maintains its dominance, but manages to win the active consent of those over whom it rules" (1989, 244). In this definition, the state, a class-based entity, encompasses the domain of civil society. In the close relationship Gramsci formulates between the state and civil society, the space NGOs inhabit becomes very important. In Gramsci's theory, it is the state's localized and everyday forms that are of the greatest importance. The state, in Gramsci's view, is a bundle of sites and scales of governance and authority that shift, grow, and continuously change. In the modern state's advanced-liberal form, NGOs can play a key role in the state's continuous construction and reconstruction, as they take on the roles the state pushes out of its purview. They can also serve, through partnering programs, as a link between the state and communities in which the state has otherwise disengaged.

Foucault, rather than focusing on what constitutes the state, uses his theory of governmentality to examine what the state does and how. In Foucault's theory of governmentality, the state is more than just the strictly political institutions and powers commonly associated with it. It also encompasses the institutions that govern more broadly by controlling and shaping everyday behavior. Governmentality encompasses the rationalities of rule as well as the forms of knowledge and expertise necessary to produce governable subjects. As Gillian Hart notes, Foucault decenters the state and brings to light the multiplicity of other sites that factor in important ways into the production of political subjects. NGOs occupy one such site (Hart 2004). In the colonias, NGOs play a significant role in what Thomas Hansen and Finn Stepputat term a key "language of governance," producing resources and ensuring the population's well-being. NGOs in the colonias help with resource acquisition and, in so doing, can become "transmitters" of neoliberal governance (2001, 7).

Together Gramsci and Foucault allow us to acknowledge both the essential political instability and partiality of the state (Gramsci) as well as its constantly reworked processes of governmentality (Foucault). Within liberalism as currently experienced, "civil society is brought into being as both distinct from political intervention," and, as such, a site for possible resistance to neoliberal politics, "and yet potentially alignable with political aspirations" (Barry, Osborne, and Rose 1996, 9). When both referenced, Foucault's and Gramsci's theories can be used to theorize civil society as a dynamic space that is not inherently liberating or oppressive but has the potential for both. The decentered advanced-liberal state, as only one aspect of the neoliberal political project, produces political subjects from many sites. Although many NGOs in the colonias attempt to use their position within civil society to work for social justice, they often become a site of neoliberal governance and end up reinforcing the very production of neoliberal political subjects they seek to deconstruct. I argue here that one central way NGOs are brought into neoliberal governance is through techniques of professionalization. The accountability, professionalism, and development of expertise that currently occupy much of the time and energy of NGOs (including those discussed here) are part and parcel of the neoliberal project. These "technical" concerns of governmentality can be just as important as ideology. They can also detract from an NGO's goal of creating or helping to instigate deep-seated progressive social change that challenges systems of power.

While the increasing presence of neoliberalism in the realm of civil society in the colonias was obvious, there was a more liberatory conception of civil society present as well. The potential for civil society as a revolutionary space was an underlying theme in the work of many NGOs. Although this liberatory space rarely materialized, it is important to acknowledge its presence, especially in relation to the current plurality of thinking on civil society. Judy Howell and Jenny Pearce distinguish between two primary theories of civil society: the mainstream and alternative approaches. We have seen in the colonias the spread of mainstream thinking about civil society, which views civil society as primarily relevant to solving the state's dilemmas and providing for the state's agendas. Yet alternative ideas about the role of civil society are also present in the colonias as activists "embrace the concept of 'civil society' as expressing a role for NGOs beyond that of implementing the donor agenda" (Howell and Pearce 2001, 16).

Following Habermas's theorization of the failure of seventeenth- and eighteenth-century bourgeois public sphere, activists and others seek to "decolonize the life world" of "personal relationships and communicative action" in order to create a more receptive space for alternative models of civil society (57). Similarly, John Keane points to the "unharnessed potential" that the alternative perspective holds to "develop new images of civil society that alter the ways in which we think about such matters as power, property . . . and violence" (1998, 190). It is firmly in the alternative perspective that, I believe, the NGOs in the colonias would position themselves. But the demands of mainstream civil society often appear to be too strong to resist.

In the colonias, NGOs encourage leaders to organize their communities around self-help projects that neatly fit into the neoliberal political project's focus on individual self-reliance. NGOs often facilitate the state's divestment from the colonias by providing private services and, at the same time, act as a transmitter of dominant forms of governance from the neoliberal state to the colonias. Deborah Martin argues, "Governance theories highlight an increasingly blurred relationship between private and public actors in public policy and service provision" (2004, 395). This blurring was an important characteristic of NGO work in the colonias.

To review, colonias were developed in relative isolation as a response to the lack of state-sponsored affordable housing during the late 1980s and early 1990s in the "roll-back" stage of neoliberalism—a period characterized by "the active destruction and discreditation of Keynesian-welfarist and social-collectivist institutions (broadly defined)" (Peck and Tickell 2002, 384). More recently, colonias have been the site of the "roll out" of neoliberalism, as colonias have been brought into the fold by both state agencies and NGOs. In turn, NGOs and the state add to the production of colonia leaders and residents as political subjects that consistently chose neoliberal solutions to their problems rather than make demands on the state. Civil society in the colonias thus "flourish[es] precisely as a consequence of state action" (Hyatt 2001, 227).

A great deal of work has been done on the changing role of civil society in developing countries and the key position of NGOs in this change (Bebbington 2000; Hintjens 1999; Mohan 2002; Morris-Suzuki 2000). In the colonias, as in many other immigrant communities across the United States, "local neoliberalisms" developed along with similar global-scale neoliberal processes and structures (Peck and Tickell 2002). A focus on the specific and concrete manifestations of neoliberalism in the colonias allows me to situate Mexican working-poor communities in the United States next to other marginalized communities within the global-scale processes and structures of neoliberalism. The story to follow exemplifies the many pitfalls that can befall NGOs and the pivotal role they play in shaping activism in the colonias.

NGOs and Change in Colonias

Many of the problems I observed while in the field revealed the large-scale difficulties that exist between NGOs and their clients. I found, just as Michael Edwards and David Hulme did over a decade ago, that problems could arise from the differences between funder goals and those of the grassroots communities that NGOs serve (Bebbington 1997; Edwards and Hulme 1996). In the worst situations, the NGOs insert their own priorities into the process, which obscure the desires of the communities. When funding priorities do not match community priorities, the former, which determines the viability of NGOs, often wins out (Townsend, Porter, and Mawdsley 2002). Because much of the time the clients of the NGOs have little or no contact with the funders, as is the case in the colonias, it is difficult or impossible for them to realize that community goals are not accurately transmitted to the donors.

If NGOs are not as progressive or democratic as they seem from the outside, a question arises: how does this complex, but not-always-progressive relationship between NGOs, civil society, and the state play out in daily life in colonias? In the colonias, NGOs like the COG, the Border Water Group (BWG), and Affordable Housing, Inc. (AHI) play an important role in the development of these communities and help initiate and lead major community projects. But they appear to have little success in terms of politicization or social transformation. While leaders like Juana, Flora, and Marie work hard to implement projects in their colonias, they rarely get their colonias as a whole to question why the communities have to live without or why the state does little to make developers build the necessary infrastructure. Discussions of their position as marginalized communities never move farther beyond the role of racial and ethnic discrimination, and most of these discussions take place in meetings limited to colonia leaders and NGO staff. Questions as to why Mexican immigrants receive worse treatment than other immigrant groups, or if other groups of immigrants encounter similar problems, rarely arise. Strategies for respond-ing to their problems or creating coalitions seldom come to the surface, and, if so, they are not followed through.

During my time in the colonias, I witnessed a series of projects initi-ated and several completed. In all, I saw similar problems play out in the relationships between the colonia leaders and the NGOs that serve them. Many of the problems I observed reveal the large-scale difficulties that exist between NGOs and their clients.

As mentioned earlier, the COG and the leaders with whom they work act as conduits of information for the colonias. In this role, the COG is often asked by local government and federal agencies to present projects to the colonias or act as a representative of a larger NGO in order to ini-tiate a project or distribute resources locally. This role can be described as that of a "broker/consultant," and many NGOs commonly assume this role (Mohan 2002, 143). This is the role the COG assumed in the housing project that AHI funded in Los Montes.

For two years, starting in November 2000, the colonia of Los Montes was the site of a housing project designed to build dozens of new homes. But, in the end, only two households, both headed by single mothers, expressed interest. The apparent failure of this much-needed project reveals the problems created by the growing emphasis on professional-ism and competition in NGO circles. It also sheds light on the continu-ing problems that internal divisions among colonias cause, especially when they emerge from competing visions for a community's future. My first experience with this project came at a COG meeting in which Elena announced that she had been contacted by the director of another local NGO, AHI, which wanted to do a housing project in the colonias in which the COG worked. AHI had put together a funding scheme that would allow colonia residents to build new homes using self-help and sweat-equity methods and to finance the projects at incredibly low interest rates. All the COG had to do was administer a survey to demonstrate the need for new housing stock, elicit interest in the colonias, and help residents complete the rather long and complicated application forms. Because the project required intensive labor from the COG, AHI had budgeted funds to pay for the part-time services of one of the COG's organizers. Like most NGOs, the COG constantly searched for more funding. The opportunity to secure more money for its payroll provided a great incentive to join the project, especially because the project would truly benefit the colonias. At this point, the project still appeared to be a good idea. Housing stock was not one of the colonias' listed priorities, but no colonia was going to complain about the opportunity to rebuild.

A meeting early on in the planning process showed signs that this project laid in the hands of the "professionals" and not the colonia leaders. At this meeting, representatives from the COG and AHI discussed the proposed project with the county planning office. No colonia residents were present, even though the project was to occur in their community. One of the county's key concerns was that Los Montes lay on a ten-year floodplain that first passed through one of the many dairies in the area (meaning the colonia flooded with cow manure), which meant the colonia should never have been built there in the first place. The county suggested that, rather than build new houses at the present location of the colonia, the money should be used to move the colonia to a better location. The COG stood up for the colonia and told the county that under no circumstances would the residents of Los Montes agree to move. Some of them had lived in that spot for more than a decade and had made the colonia their home. This was, unfortunately, the first and last time anyone really represented the colonia in meetings on this project. Once the county agreed to the project, with the stipulation that the new buildings be built more than a foot higher than the existing structures to avoid flooding, the COG could proceed with the survey.

The COG staff and I administered the survey. None of the colonia leaders were asked to help, even though the COG staff told them to be home the day of the administration. At this point, talking to their neighbors was the only role colonia leaders had assumed in this project. On several occasions, Flora asked me for information about the housing project. I answered as best as I could and then asked the COG organizer in her community to provide the information. Flora and the other leaders were on the outside of this project. In the long run, their absence spelled failure for the project. Flora and Esperanza, both leaders, were the only two residents to sign onto the housing project. It is no coincidence that they were both the heads of single mother households. The same daily practices that made it easier for women from female-headed households to become leaders made it easier for Flora and Esperanza to sign on to the housing project. The families headed by married couples all declined to apply for the project for the same reason: the women were all interested in the project, but their husbands said they could build the house themselves and that they did not want to go further into debt. Their wives expressed doubt that this would ever happen, but they nonetheless did not sign on.

The fact that most of the COG's background research on community support for the project was done through Flora and Esperanza requires closer examination. As single mothers and women heads of household, their points of view on the project, and community needs in general, were often different from those of the women and men from male-headed households. Male-headed households did not hold new homes as high on their list of community improvements as women in female-headed households did. Had there been more leaders from male-headed households, or simply more community participation in the project, perhaps the COG would have seen the split in the colonia's perception of the project and its importance to the community. But as it stood, with the limited level of community-wide participation, the competing visions of various colonia households were not clear to the COG, and the project did not succeed as planned.

Seven years later, in the summer of 2007, only two new houses from this project stood in Los Montes. The importance of community support for a project is clear. The COG signed onto the housing project in part because it would raise money for its payroll. If it were successful, the project would have produced very obvious and quantifiable results. But the project failed to produce many new houses or create long-term revenues for the COG, who had to end its contract with AHI because of the lack of interest in the colonias.

When funding for house construction is all that is available, many NGOs, such as the COG, will divert their activities to accommodate funding trends. In the case of the housing project in Los Montes, this new focus not only overshadowed community priorities but also led the COG into a very complex situation they were poorly prepared to address. In its efforts to present itself as the colonia expert, the COG missed crucial signs that would have told them the project was on shaky ground. The COG was caught up in a cycle of professionalism. They were carried away with their eagerness to complete a large-scale project, and they lost sight of the realities of colonia life, such as the fact that projects that involve little or no financial burden are most popular with men and that, without the support of the colonia's men, more costly projects become impossible. Had Flora and Esperanza talked to the colonia about the project from the start, they might have been able to identify and resolve the discrepancy between the women's excitement and the men's reticence about the project. Yet, when NGO employees did discuss the project with leaders, the lack of male support was not apparent. NGOs are forced to work within a system that puts so many demands on their time and resources that it becomes very easy to lose sight of the community goals, community dynamics, and the daily politics of community organizing. Misunderstandings and misinterpretations of community dynamics are a constant problem for NGOs. This problem is often intensified, as it was here, by the increasing pressures that outside organizations, such as donors, put on NGO operations. Part of the professionalization process is the development of continuing sources of funding, and this increased focus on funding can be detrimental to the fulfillment of client needs.

The two houses that were built represent a vast improvement in the lives of two families. But the relationships that NGOs develop with communities produce more than material improvements. These relationships can also justify and reinforce the unequal power relations that led to these poorly planned communities in the first place. Janet Townsend Gina Porter and Emma Mawdsley found the international aid system to be a "chain of dependency-inducing relationships" and that those "partnerships do not emerge in a vacuum. They emerge from an existing institutional architecture" (Ling 2000 in Townsend, Porter, and Mawdsley 2002, 834). In colonias, NGOs focus on short-term, immediate projects as opposed to long-term structural changes that potentially politicize not only colonia leaders but also communities, who then work to change the processes that create their own marginalization. It is the crucial processes and relationships of marginalization that many NGOs strive for yet ultimately fail to illuminate. NGOs can actually silence the already-marginalized voices of colonia residents. David Hammack theorizes the historical role of NGOs in moving social conflict out of the open and into more controlled outlets, such as community organizing (2002). In the colonias, NGOs inadvertently do this by situating activism into the realm of the individual.

The housing project attracted the COG on several levels. First, the obvious need existed. Next, AHI could offer funds to cover payrolls and daily costs. Finally, the project promised concrete and easily accountable results. The COG is quite aware that houses are perhaps the best indicators of a successful community-development program. Rose, Fisher, Mohan, and others discuss the growing importance of accountability as NGOs become more and more professional (Fisher 1997; Mohan 2002;

Rose 1999). During my year in the field, annual reports and donor surveys constantly plagued the COG's director. She spent more time juggling donor and community agendas and writing grant reports than she ever spent in the communities. An NGO's accountability can be moved from clients to funders very quickly, posing a real problem. Elena is still torn between her desire to be on the ground in the field and at her "place" behind a desk in the office, the place where colleagues tell her she belongs as the executive director.

While I was in New Mexico, the COG held three fundraisers and wrote an entire new position into their budget just to pay for a full-time grant writer. Procuring funds can easily become the number one preoccupation of NGO directors. This focus on money leads many NGOs to put as much effort into fundraising as they put into program implementation. With the incredibly high levels of competition for funds, it becomes even more necessary to put in time on follow-up reports and yearly statements. The main concern of these reports is accountability; how can the NGO show the donor that it has done what it said it would do with the money and argue that it deserves another year's funding? In this process of creating accountability, it is not uncommon for an NGO to describe its work in terms it would not normally use but that exist in the vocabulary of the funder. NGOs often rework the focus of their services to please funders. According to Mohan, NGOs "are so wary of upsetting their funders that they tightly circumscribe the activities on the ground and completely undermine independent development. Their partners [NGOs] are trapped in an irreconcilable position of being the authentic representatives of their grassroots constituencies, but being accountable to organizations outside the locality. Squeezed in such a way they usually defer to the funder and present to them a relatively trouble-free view of local communities and their development needs, all of which further marginalizes and alienates the rural poor" (2002, 148).

In a move to help the colonias create more lively economies within their communities, the COG hired Sarah as an economic development program coordinator. Sarah spent the better part of her first year finding funds and looking for projects that could be carried out in the colonias. After careful research, day-care centers appeared to be one of the best options. They provided a much-needed service the state was cutting at every turn in its efforts to downsize, and one that working families desperately required. At the same time, they provided employment in the community to women who could only work in or near their homes while they watched their own children. At least this was the initial reasoning. By the time Sarah was comfortably established in her job, the COG had gone full tilt into the day-care business. At this time, the COG ran a day care in one colonia and was starting two more day-care projects in other colonias. Once Sarah was deeply involved in the day-care centers and had written several successful grant proposals, her time became fairly well locked into the centers. Other economic development projects, such as a *tortilleria*, a small-scale farming business, and an animal-husbandry project, that other colonia residents wanted had to wait. In order to acquire funding, Sarah had to become an expert on day-care centers. Although basic infrastructure may be the highest priority for most colonia residents, Sarah has to present economic development as a high priority to the funders, which could downplay the extreme poverty in which colonia residents lived. The very fact that "economic development" is a funding priority for so many NGOs and donor organizations speaks to the ways in which donors can shape the goals of NGOs and, in turn, the services these NGOs offer to their client communities. NGOs must make themselves competitive in order to access funds. This can mean adopting projects that are fundable based on the current trends in community development. In the end, it became clear that day-care centers were to be only a limited success in the short term in the colonias because they were mired in regulations from construction to daily operations, making them very difficult for colonia residents to run. The COG could not simply rent a space and start a center; massive remolding was usually required, and then a certified person was necessary at all times to supervise the center. Most colonia residents did not have the necessary background for this position because they did not have the education or training. Once these obstacles were overcome, many of the women who were initially interested in working in the center had moved on, and Sarah had to recruit a new group of women. For all the day-care centers Sarah finally got off the ground, as many discourses of caretaking and motherhood were used to solicit both paid and volunteer staff as those used to recruit community leaders.

Not only are NGOs expected to be experts and professionals but they are also expected to compete like full-blooded capitalists—even though they are usually not supposed to make a profit. The stress on quantifiable concrete and practical changes in individual communities that fuels this competitive system can limit the development of politicization and empowerment in colonias by obscuring the ties between marginalized communities. Mohan sees "fragmented politics" as the result of this stress on individual projects (2002). Others describe this situation more forcefully as "de-politicization" (Fisher 1997; Hintjens 1999; Manji 1998).

Whether one calls it fragmented politics or depoliticization, the results are the same. NGOs are being forced to transform themselves into market-competitive organizations in order to get funding. To compete they must present themselves as experts, create a niche, and defend their territory. The COG's primary grant writer, Marlene, said she had to be "always on message, on our story, self-conscious of our narrative," and if they were, funders "would buy the narrative" and give them the money: "Trying to not only do our work, but also sell ourselves, is a very hard thing to do and it takes a lot of energy."[2] As NGOs become more entrenched in this system, they are less likely to serve the desires and needs of their clients. They are certainly less likely to see through to the often contradictory desires of the communities themselves, such as the gendered differences in views on the homes-building project in Los Montes or the competing visions of leaders as seen in the split between Flora and Lola.

Marlene informed me that on two occasions she had jumped at a grant, wrote a very persuasive application, got the money, and then had to sell the project to the staff who were less than excited about the idea. In her examples, funding priorities were twice removed from the communities and were clearly donor driven. More profoundly, in these situations, the ability of NGOs to politicize client communities shows signs of diminishing, and the networks NGOs have worked so hard to create are less likely to bring marginal communities together to examine the processes of their own marginalization.

The confusion that shrouded the housing project was due in part to faulty communication between AHI and the COG. Rather than enter Los Montes on its own and organize the project itself, AHI went through the COG for two main reasons. First, AHI has an ongoing relationship with the COG. The director of AHI and Elena, the COG's director, are friends. In order not to step on the COG's feet, AHI proposed the project as a joint effort. Second, by working through the COG, AHI could offer to fund some of the COG's work in the colonias by paying an organizer to work on the housing project. Finally, the COG describes itself as the "gatekeeper" of the colonias and has a reputation for fiercely protecting its interests in the colonias. When other NGOs like AHI want to start a project in a COG colonia, it is in their best interest to go through the COG.

This role of "gatekeeper" to the colonias is a problematic one for the COG. In my experience, the COG employs the term to explain their role in the colonias as that of a protector who makes sure researchers or other NGOs do not constantly bother the colonias' residents. When I first presented my project to the COG, it clearly stated that I would need its approval to initiate the study and that I would have to play by its rules. By and large, I believe its rules are valuable and that they are in the best interests of the colonias, yet I cannot help but view these rules and the COG's role as "gatekeeper" as rather paternalistic. Mohan found similar paternalistic relations between NGOs and grassroots communities: "The SNGOs [Southern NGOs] are taking ownership of the local culture and using it as a defense mechanism. The NNGOs [Northern NGOs] realised they needed to have intermediaries, ideally working in a partnership relation, but the SNGOs use this powerful position to protect their constituency of villages. They claim to represent the local communities, but have rather patronising attitudes towards them, but know they are beyond reproach. In this way civil society organisations actually impede democratisation and good governance" (Mohan 2002, 143). If we take SNGO to stand in for the on-the-ground NGOs in the colonias and NNGO to stand in for the larger nonprofit and federal donor organizations like the Ford Foundation, Heifer International, HUD, or the Environmental Protection Agency, then we can see how this quote describes a situation in the global south that works just as well on the U.S.–Mexico border, a boundary between the global north and south. In the colonias, the relationship that develops between funders and community groups can be seen as patronizing and can also be described as "partnering," which certainly has patronizing elements as well. This patronizing relationship between the COG and the colonias is only exacerbated by the competitive funding system that pits NGO against NGO. NGOs like the COG are expected to be gatekeepers, protectors, experts, and brokers in their relationships with their clients.

This intra-NGO competition is tied to the requirements of funders, which force NGOs to work on tangible projects whose results can be measured and reported. It is now clear that the NGOs that serve the colonias are not that far removed from the state and are certainly part of the economic system the state supports, that is, neoliberal capitalism. In this system, even NGOs must be market competitive, which means the

COG needs to protect its investment in community organizing by keeping other NGOs out of its territory. Just as the neoliberal political project has deregulated many previously public services in the belief that market forces and competition will lead to better services and lower prices, NGOs are encouraged to compete for funds in a capitalist manner by many federal agencies in order to create better options for their clients. So the question becomes, how can this be avoided? NGOs need to find a cooperative space outside the market competition of capitalism, a space to share projects and communities, and relate seemingly disparate projects. The colonias and their leaders need to see how the wastewater project is related to the housing project and how both are part of the lack of county support and are tied in basic ways to the neoliberal political project. As Tessa Morris-Suzuki points out, the intricate connections between the material needs of communities and their economic and political marginalization are all joined together (2000). The only way these connections can be made is for the COG and other NGOs, such as the BWG, to work together and make their work transparent to the communities. One way might be to "scale up" local interventions in an effort to clarify the politics of community work (Mohan 2002). Through a "scaling-up" effort, colonia projects and issues can be linked to larger scale struggles for equal representation, economic justice, and gender- and sexuality-based equality movements, among others.

The housing project was a self-help project in which colonia residents worked side by side with professionals to defray coats. This project, like other self-help projects, did not appear to empower leaders in the colonias or contribute to substantive political change at the scale of the community. Why is this the case? To start with, social change is a process, and, for self-help projects to form part of this process, there needs to be some kind of continuity between projects and the other interventions of local NGOs. Here lies one of the crucial impediments in this process: the competition between NGOs. Most of the work done on NGOs and their relationships to each other and their donors comes from case studies in international development situations. Nonetheless, much of this work applies equally well to the relationships between NGOs and clients that I observed in the colonias. As NGOs grow in importance in the United States, the "NGO industry" looks more and more like its predecessors in the global south and begins to experience similar problems as those discussed in the development literature.

Conclusion

The problematic relationship between NGOs and their clients as demonstrated by the relationships between colonia leaders, the COG, the BWG, and AHI have repercussions beyond the unfinished projects and limited politicization of these communities. By focusing the attention of colonia leaders on the local scale and keeping most of the politics of community organizing to themselves, NGOs also limit the progressive social change that colonia leaders could create to confront systems of power.

It should be abundantly clear by now that global economic restructuring and neoliberal policies have led women to take greater responsibility in the processes of social reproduction and that NGOs play an important part in this process by providing technical and organizational support. But NGOs have another, even more important and problematic job: they tie colonia residents to the neoliberal state as part of the neoliberal political project through their position in civil society. Many NGOs inadvertently constrain colonias in ways that benefit the state and reinforce dominant and marginalizing forms of social reproduction.

Most NGO interventions in the colonias try to alleviate practical problems. Many have material results that can be tied to the daily concrete practices of social reproduction: for example, the installation of potable-water systems and natural-gas lines. But that is not all NGOs do; they also have an important discursive role in their position as mediators between the colonias and the state. In this way, NGOs, inadvertently or not, often propagate dominant neoliberal discourses on leadership and activism, producing a particular kind of subject. By influencing projects and community organizing, NGOs can govern communities in ways that help the neoliberal political project by creating self-reliant citizens who provide their own basic services and support Rose's "advanced liberalism" (1999).

This research suggests that NGOs' failure to foster alternative forms of daily social reproduction rests, in part, on the landscape in which they must work. NGOs are the mercy of the neoliberal political system, as they work to help those that are most imposed upon by the very same system. Through their position in the "shadow state," NGOs like the COG must comply with both their funders' priorities (often based on state trends in funding) and the state's regulations and agencies (e.g., local housing codes and HUD). In this playing field, designed and managed by others, it is no wonder that NGOs have trouble meeting their goals.

In the production of counterhegemonies, I see the potential of NGOs to create politicization, empowerment, and progressive, radical social change. Civil society links daily material life with the discourses and processes of the neoliberal political project through the project's roles in service provision and the maintenance of the dominant cultural hegemony. In order to use this role to facilitate change, NGOs must understand their positions in civil society and as bridges between the neoliberal order and possible alternative orders.

Epilogue

Seven Years Later

On a hot summer evening, we walked along the edge of a cotton field and then down another hard-packed dirt road that skirted the back side of someone's lot. As I got that familiar feeling of almost trespassing that one has nearly all the time in a colonia where many lots simply bleed into each other and the street, Flora explained that there was a very old señor living back here in a tiny trailer under a tree. Well, his mattress was under the tree, there was not enough room in the one-room travel trailer, the kind usually pulled by a truck, for a bed and a table. While we were out walking, she explained that she wanted to take the opportunity to check up on him. At this point we had walked nearly the entire length of Los Montes, and Flora knew pretty much every lot and every family in the colonia. She told me she had met this gentleman when she went door-to-door earlier that year to get signatures for the new road project. He had lost his home and was squatting on a shady bit of land under a tree. The owner of the lot said it was OK for him to stay there, but, if the county found him living out in the open like that, they would certainly make him move out and into a shelter. For an older man like that, losing his independence would be crushing. Flora knew this, so she tried to check on him and help him stay where he was without drawing any attention to him. As we came around the hedge and his "home" came into view, it was a slightly different scene for a colonia. It was not the usual rundown mobile home; it was more of a campsite with a mattress and a chair set up under a big shade tree, and it was really very peaceful. Yet like many other colonia homes, it only held the bare minimum. And unlike the typical campsite, this would not be packed up anytime soon, and he would not be going home to a hot shower.

As a leader, Flora is an activist, an organizer, a community builder, and a friend. She is also a social commentator and an astute one at that. She exemplifies the politicization possible, but not common, on the individual scale for community leaders in the colonias, a politicization she herself acknowledges does not exist at the community scale.

How We Got Here

This project, like many others, began as one thing and turned into another. When I left for New Mexico for the first time, my primary concern was to discern why it was women and not men that took positions of leadership in the colonias and what role nongovernmental organizations (NGOs) had in this process. Once I had spent time in the colonias with the leaders, it became clear there was no single process by which women became leaders. It was much more complicated than that, and NGOs were by no means the only actors in the web of relationships that led certain women to take on leadership roles in their colonias. I also began to realize with time, and much of this hit me after my initial fieldwork, that these amazingly powerful women who held so much potential to politicize their communities were caught in a system that undermined their work. The same, of course, had to be said of the NGOs. This was a deeply disturbing realization at first. I was disappointed because I had seen the work of NGOs and colonia leaders as entirely positive up until that point. But then I realized that I had to see this as an opening, an opportunity, to contribute something to the communities in which I had done this research. So far very little had been asked of me by the communities themselves, and I was looking for ways my work could be used beyond the basic reports to funders. I saw the opportunity to examine this discrepancy and produce a possible explanation as a way to give the communities and NGOs something very different and possibly valuable and also create a study of a highly disabling process.

As we have seen, in the colonias, the nearly ceaseless efforts of women leaders with the help of NGOs create much-needed infrastructure. In this process, some women leaders become politicized, but the vast majority of colonia residents show little interest in community organizing or political projects. When we focus closely on the type of political subjects being produced in the colonias, it is less surprising that residents remain detached or individually centered in their activities in their community. We find tremendous outside pressures on both NGOs and women leaders that shape women's activism and leadership to create neoliberal forms of community through self-help techniques and neoliberal forms of government. Within this individually centered neoliberal framework, politicized activism is simply not the most obvious choice for many people. If leaders do not spend time analyzing their own oppression and marginalization, neoliberal rhetoric of self-help and self-reliance sounds awfully good. In the colonias, it was more than leaders and NGOs could combat on their own.

Even on the level of leadership itself, I soon realized that it was the very idea of motherhood under which women mobilized that was being co-opted by the partnering state and NGOs to recruit women as coworkers and organizers in community-development programs. This is only one of the many ways the neoliberal political project has worked its way into the daily lives of colonia residents, turning these immigrant communities into examples of the perfect, self-reliant, neoliberal communities that provide for themselves what other nonimmigrant, not-of-color, non–working poor communities have provided for them.

In explaining this shift from activist-leaders who produce community-wide resistance to activist-leaders who primarily educate themselves and "fix" the community, I wanted to shed light on the dirty work of the neoliberal state at the behest of the neoliberal political project as it steps into the third sector of NGOs and community organizing. The unfortunate end result is the disabling of activism on the community scale. Much to my dismay, the primary mechanism for much of this process are NGO interventions made with state funds, on state-based initiatives, or simply based on state-influenced trends. The very same progressive NGOs that trained these breathtakingly brave and powerful women end up undermining their advancement and often that of their colonias.

I have attempted to demonstrate how complex the relationships are that construct leadership in the colonias and why it is crucial to acknowledge this complexity. Hegemonies, even the colossal neoliberal hegemony, are never complete, and it is in their gaps and openings that change and revolutions can occur.

Clyde Woods, in a discussion of communities of color in the United States that have essentially been abandoned to be destroyed, asks, "Are these communities fragmented because they are on the margin of civilization or is it because they are on the front lines of globalization and a global racialization" (2002, 63)? I believe the same question can be asked of the colonias and that the answer tells us a good deal about the role women leaders are asked to play in their communities and why the task they take on is so difficult. In the colonias, I believe the answer to this question is both—the colonias are fragmented physically, socially, and culturally as transnational communities because they are both marginalized at the edges of what most people consider habitable living conditions and because they have been obviously and purposely created by economic integration between the United States and Mexico. Colonias are the direct descendants of the Bracero program; they are the physical manifestation of the North American Free Trade Agreement and Mexican immigration to the United States for employment. If that is not "the front lines of globalization," then what is? Colonias are certainly racialized spaces and have been marginalized because of this for decades. Woods suggests the importance of regional studies of the human rights violations of repressed groups in order to understand "the weight of history" (64). That is what I have tried to do: to use the "weight of history" to situate the current state of politicization, or lack thereof, in the colonias.

What Now?

If politicization is the goal, as it is for many of the NGOs that work in the colonias, than how do we get there? Perhaps first we need to think about why this is the goal, and, more to the point, whether this is a realistic and beneficial goal and for whom is this goal realistic and beneficial? As I have worked on this project, given talks, and written papers, I have often been asked, "Why politicize?" Why not just use self-help methods and get on with it? Why should colonia leaders bother with creating progressive social change on the community scale and why does it matter if their relationships with NGOs actually disable their ability to spread political awareness? These questions have certainly made me think, and at times I have wondered myself, "Am I just imposing my own desire for progressive social justice on these communities?" After all, I will be the first to admit that there are many factors at play in the life of a colonia family, and adding more pressure in the form of activism to the life of a person who is working to just stay afloat seems unfair. Yet again and again I feel myself pulled back to the words of the women with whom I worked and their deep seated desire to pass down a better life to their children and their community. I know that this better life has to be one in which colonias are not produced as marginalized spaces, and the only way to do this is through the work of politicized residents who demand what is required to make these communities safe and healthy. Yes, the same infrastructure can be created through self-help techniques, but the reproduction of colonias as resident-defined, -produced, and -politicized spaces must be done by holding the state and developers responsible for the abandonment of these communities. Once this is done, the changes that currently are being made because of the hard work of a select group of women leaders could be spread across all the colonias. Just because Margaret Thatcher so firmly believed "there is no alternative" to neoliberalism does not mean colonia residents have to buy into her delusion. For colonias to move beyond their production as neoliberal spaces, they need to also move beyond their entrenchment in neoliberal ideology. This is a major structural change that may not be realistic immediately and will certainly have to happen on multiple fronts where the discourses that have produced colonias as neoliberal spaces originated. Discourses can be actively manipulated in friendly agencies like NGOs, but, at the state level, it will be harder. These discourses will require active deconstruction. Once the atomized neoliberal identity can be broken down, the next step will be to politicize—to move beyond the neoliberal agenda.

Yet the question still remains: how to politicize beyond the individual women leaders? Is it clearly up to the NGOs and women leaders to decide what is next? They may not have come here to "fuck shit up," as Ernie so succinctly put it, but they also did not move into the colonias to live in spaces socially constructed as waste dumps or physically produced and used as such. It certainly is not up to one or two women in each colonia to organize the community to overcome these obstacles because then we are back where we started. The problem appears to be in the numbers. This was always the issue when any NGO went into the colonias looking to do work. Getting residents interested was easy at first, but keeping them interested and coming back to meetings and involved over any period of time was always difficult, if not impossible, as we saw in the organizing of Valle. As community organizers know, "life happens," and people stop coming to meetings. But this means you are relying on the same women over and over again, and this is problematic, especially so when these women are recruited through discourses that produce them as leaders based on their qualities as mothers and caretakers.

We need new discourses that produce leaders based on different, more egalitarian qualities with which both men and women, married and single, mothers and otherwise can all be leaders and activists in the colonias. Once we admit that the neoliberal political project is not inevitable, we can see that there are alternatives to the neoliberal hegemony and that, particularly in marginalized communities, discourses of identity, self-reliance, and self-sufficiency rob communities of the social cohesion and collective work that community organizing can bring. Wastewater systems and paved roads, although absolutely necessary in the colonias, are just the beginning of what an organized colonia can accomplish. On my last visit to Los Montes, Flora and I passed a sign for a community-based civil rights network that documents abuses of the border patrol in the colonias (see Figure E.1). She commented that if more people were involved in the work of that organization, *La Migra* would not be able to harass colonia communities with impunity. Seven years after the training in Del Cerro, which Flora attended, harassment by *La Migra* is still a big problem in the colonias, and very few residents are involved in work to document and challenge illegal raids. This is the sort of politicized organizing in which colonias can take part when a more collective social milieu exists. If the neoliberal ideology of "each person for herself" were lifted, the whole community could step up and take the lead in improving conditions and talking back to stereotypes of colonia communities.

In this present moment of danger (Pred 2000), when the rights of immigrant communities in this country are under attack, the politicization of these communities is more important than ever. As I am writing this, my e-mail inbox is overflowing with messages from a "border news" listserv on the current surge in arrests and raids on both undocumented and documented workers and residents across the United States. A recent series of these messages hit home harder than usual for me. These headlines spoke the names I had been uncomfortably waiting to hear. I knew the U.S. Immigration and Customs Enforcement (ICE) would not stay clear of the colonias in New Mexico forever. I knew I would wake up one day and find an e-mail from a colleague at an NGO or on the listserve describing a raid on a colonia, and I know, someday, someone I know will have their life destroyed by this process. This time I was spared in a small way; the colonia raided was not one in which I did much research and, in a very selfish way, I was relieved that most likely no one I knew would be impacted. Of course, the NGOs with which I worked would be devastated. I have been following the growing anti-immigrant sentiments in the United States since the 1990s but have found the irony and severity of the latest round of immigration legislation debates and interventions since early 2007 shocking nonetheless. As I have mentioned before, the majority of colonia residents are documented, yet their communities are regularly targeted by ICE, and now they also appear to be the target of these new and improved mass-deportation raids. They are meant to be

Figure E.1. A sign posted in the front yard of a lot in Los Montes. It reads, "Regional Citizens' Council, Regional Documentation Campaign. Report Abuses by the Border Patrol, Police, Customs, and Other Authorities."

specifically targeted at populations produced discursively as those taking advantage of the system. How can the residents of a community lacking basic services be taking advantage of the system that requires it to supply its own infrastructure? This system gives them nothing of which they might take advantage.

In a society where the spaces in which immigrants can feel safe are shrinking daily, and as these groups are demonized by the media and the government, politicization is going to be more and more important for creating sustainable immigrant communities. Colonia residents, activists, and leaders will have to keep in mind the structural conditions under which they work when developing strategies for the continuing survival and growth of their communities. Knowledge of the obstacles to organizing and the discourses and powers that affect the formation of political subjectivities and the production of leaders will make the creation of united, politically aware, and safe colonia spaces easier.

Notes

Introduction

1. "Facts about Farmworkers and Colonias," U.S. Department of Housing and Urban Development, http://www.hud.gov/groups/farmwkercolonia.cfm (accessed June 25, 2008).

2. I would like to counter this report to say the colonias in which I did this research were 95 to 99 percent Mexican immigrant, not Mexican American. This is a significant ethnic difference in cultural, economic, and political terms. The women with whom I worked did not want to be identified as Mexican American. They preferred to be identified as Mexican and keeping their ethnic identity tied to Mexico was very important to them.

3. Federal Reserve Bank of Dallas, Office of Community Affairs, "Texas Colonias: A Thumbnail Sketch of Conditions, Issues, Challenges and Opportunities," Texas Secretary of State, http://www.sos.state.tx.us/border/colonias/faqs.shtml (accessed September 23, 2008).

4. Personal communication with Espy Holguin, HUD Las Cruces, New Mexico, 2000. I have tried to locate newer numbers, and in more recent conversations with Ms. Holguin in 2007, I was informed that no such data exists.

5. President Bush was the governor of Texas before he was the president of the United States, and Texas has more than two thousand colonias. When he ran for president for the first time, there was a great deal of excitement in colonia circles and hope that perhaps it was finally time for the nation to become aware of these forgotten communities. But no one was terribly surprised when conditions remained essentially the same in the colonias, and, if anything, life on the border became more and more difficult during the two Bush administrations in 2000 and 2004 due to severe immigration legislation.

6. "State and County Quick Facts," U.S. Census Bureau, http://quickfacts.census.gov/qfd/states/35/35013.html (accessed March 22, 2008).

7. "Quick Tables," U.S. Census Bureau, http://factfinder.census.gov/servlet/QTTable?_bm=n&_lang=en&qr_name=DEC_2000_SF1_U_DP1&ds_name=DEC_2000_SF1_U&geo_id=05000US35013 (accessed March 22, 2008).

8. Ibid.

9. "State and County Quick Facts," U.S. Census Bureau, http://quickfacts.census .gov/qfd/states/35/35013.html (accessed March 22, 2008).

10. "Quick Tables," U.S. Census Bureau, http://factfinder.census.gov/servlet/ QTTable?_bm=n&_lang=en&qr_name=DEC_2000_SF1_U_DP1&ds_name =DEC_2000_SF1_U&geo_id=05000US35013 (accessed March 22, 2008).

11. "State and County Quick Facts," U.S. Census Bureau, http://quickfacts .census.gov/qfd/states/35/35013.html (accessed March 22, 2008).

12. In the late 1970s, Mexico discovered the extent of its vast oil reserves and began down the dangerous "oil boom" other countries in the global south, such as Venezuela and Nigeria, had gone before it (Riding 1989). As the Mexican economy grew faster than it could then finance itself, it had to borrow money to address its increasing deficit. As signs began to pile up all pointing to imminent demise, President López Portillo ignored them and focused instead on creating jobs and waiting it out. When López Portillo's successor De La Madrid took over in 1981, the country was at the edge of collapse and foreign banks began to demand repayment of their loans. All Mexico could do was close their foreign exchange markets and suspend payments on their over $80 billion in debt. Help came quickly in the form of an IMF intervention, paving the way for similar interventions in the global south for years to come (Riding 1989, 149; Wynia 1990).

13. "Mexico and Remittances," The Multilateral Investment Fund, http:// www.iadb.org/mif/remittances/lac/remesas_me.cfm (accessed April 11, 2008).

14. "Facts about Farmworkers and Colonias," U.S. Department of Housing and Urban Development, http://www.hud.gov/groups/farmwkercolonia.cfm (accessed July 18, 2007).

15. The tone of Estella's voice was just as important here as her choice of words. *Cabrona* is a rather harsh word to use to describe a woman and certainly not one thrown around lightly. But what really made the moment memorable was Estella's trademark deadpan delivery: you could never tell if she was joking or completely serious, and it took me nearly my whole year in New Mexico to be able to read her. At that point, I was working on blind faith that she was complimenting me and that it was a good idea to go along with it.

16. The theories of scholars such as Sayer (1991) and Massey (1991) can help clarify the constitutive relationship between the "local" and the "global" in the colonias. Sayer (1991) developed a typology that helps debunk the idea that locality studies are merely descriptive of unique and highly contextual situations that cannot speak to regional or transnational processes. Sayer's central point is that the assumed contrasts between dichotomies such as global–local and independent–interdependent "either break down or involve more complex relationships than is commonly realised" (283). Sayer points out that "some dualisms turn out to have a missing middle term, or to be a continua rather than dichotomies" (285). Through the study of three specific, independent, and unique colonias, this research examines

the individual daily practices, and the meanings attached to these practices, of eight women leaders in order to clarify the multiple connections between the daily activism of these women, the NGOs that help them, and the neoliberal state.

1. The Production of Colonias as Neoliberal Spaces

1. In 1845, the United States annexed Texas, making it a territory of the United States and ignoring the Mexican claim to this land. The next year the U.S. Congress declared war on Mexico in an effort to extend President Polk's "manifest destiny." The end result of the U.S–Mexican war was the Treaty of Guadalupe Hidalgo in February 1848. The treaty ensured that New Mexico, Arizona, and California north of the thirty-second parallel would become U.S. territories for the price of fifteen million dollars. In 1854, the Gadsden Purchase bought the southern parts of New Mexico and Arizona for the price of ten million dollars. Two of the colonias in this study are located in the lands purchased by Gadsden in 1854 (Owen 1999).

2. The similarity to current proimmigration arguments posed by growers is striking.

3. One of the most hotly debated items on the agenda today in U.S.–Mexico relations is possibility of a new Bracero-type temporary labor agreement. Former President Fox and President Bush discussed the idea on several occasions but did not come to any agreement, in part because of the controversy surrounding the topic in the United States.

4. Interview Estella, Recuerdos, New Mexico, August 25, 2000.

5. Interview Estella, Recuerdos, New Mexico, August 25, 2000.

6. "Maquiladoras at a Glance," Corp Watch: holding corporations accountable, http://www.corpwatch.org/article.php?id=1528#map (accessed September 20, 2008).

7. Interview Marie, Valle de Vacas, New Mexico, June 12, 2000.

8. Interview Mario, Las Cruces, New Mexico, December 5, 2000.

9. The INS is now under the Department of Homeland Security and is called the U.S. Citizen and Immigration Services.

10. Interview Mario, December 5, 2000.

11. In much the same way, Ward discusses the development of colonias in Texas after the end of the Bracero program when growers became less likely to provide housing to their no-longer-legal immigrant work force (Ward 1999).

12. The historic capital of New Mexico, Mesilla, located in Doña Ana County, was home to the outlaw Billy the Kid for many years.

13. Interview no. 1, Frank Weissbarth, Las Cruces, New Mexico, October 5, 2000.

14. Interview no. 1, Frank Weissbarth, October 5, 2000.

15. Interview no. 1, Frank Weissbarth, October 5, 2000.

16. The subdivision laws were only changed after a heated and drawn-out battle. Those that were in favor of tougher laws were colonia activists and state officials who worked in the colonias. Many ranchers and other landowners were against stricter laws because they wanted to be able to subdivide their land among their children as they saw fit.

17. Doña Ana County Subdivision Regulations, ordinance number 166-96.

18. Interview with Assistant County District Attorney Karen Acosta, Las Cruces, New Mexico, July 14, 2000.

19. Interview Flora, Los Montes, New Mexico, July 14, 2000.

20. Interview Esperanza, Los Montes, New Mexico, September 11, 2000.

21. Interview with Karen Acosta, July 14, 2000.

22. Ibid.

23. I use the term colonias to name unplanned subdivisions on both sides of the border. These communities are officially called colonias in the U.S. and Mexico. Those in the U.S. were given the name colonias in part to reflect their similarities with those in Mexico.

24. E-mail communication Sylvia, June 2000.

25. Early on in my discussion with all the women in my sample, I asked each woman how they identified themselves ethnically. All the women, with the exception of Alicia, identified themselves as Mexicana. Alicia, who was born in Texas and was raised bilingual, identified herself as Mexican American. All women described the colonias as Mexican spaces and their fellow residents as Mexicanos. There was a clear distinction in and around colonia communities between Latinos from Mexico. There were those that were "Mexicanos" and those that were "Mexican Americans"; most noncolonia Latino residents I saw visit the colonias were also Mexicanos. Although we discussed racism on the community scale and in the presidential election, we rarely discussed race in terms of their ethnic identity, and it was not a topic the women brought up in conversation, so my data here are limited.

26. There is currently a debate in colonia research around the validity of using the term "colonia" to describe communities that lack infrastructure away from the border. The argument of this work, and this chapter in particular, does not support this claim. Colonias are Mexican communities on the border, and the term "colonia" developed to describe this particular kind of community. It would be inappropriate and misleading to apply the term to other communities. It is my opinion that another term needs to be developed to describe similar communities away from the border.

27. E-mail communication Sylvia, June 2000.

28. Interview Marie, Valle de Vacas, New Mexico, November 29, 2000.

29. Interview Espernaza, Los Montes, New Mexico, December 6, 2000.

30. Interview Espernaza, December 6, 2000.

31. Interview Estella, Recuerdos, New Mexico, August 25, 2000.

32. Interview Alicia, Valle de Vacas, New Mexico, December 12, 2000.

33. Interview Juana, Recuerdos, New Mexico, September 14, 2000.

2. The Production of Women as Neoliberal Leaders

1. See the following discussion of "tradition."

2. Chapter 4 discusses in detail the complex and "messy" relationship between discourses of gender in the home and activism in the community.

3. The questions I used in this interview were very straightforward. I acknowledge that they might have elicited less complex answers than more nuanced questions would have. But what I wanted to get at was the rather stereotypical answers that make up the discourses I observed the women employing. These questions also served as a launching point for longer discussions.

4. Interview no. 4, Juana, Recuerdos, New Mexico, December 7, 2000.

5. Interview no. 4, Alicia, Valle de Vacas, New Mexico, December 12, 2000.

6. Interview no. 4, Marie, Valle de Vacas, New Mexico, November 29, 2000.

7. Ibid.

8. Interview no. 4, Juana, Recuerdos, New Mexico, December 7, 2000.

9. Interview no. 4, Flora, Los Montes, New Mexico, December 6, 2000.

10. I recently had coffee with Flora and Sara while they visited Los Angeles for a national conference on families and social justice. Sara was happy to tell me that she was attending university at New Mexico State. She was not starting a family yet, as she would like to travel before that. Paris is first on her list.

11. Field notes, Flora, Los Montes, New Mexico. June 19, 2001.

12. Ibid.

13. Ibid.

14. Most Rev. Ricardo Ramírez, C.S.B., Speaking the Unspeakable: Pastoral Letter on Domestic Violence, Roman Catholic Diocese of Las Cruces, December 14, 2009, http://www.dioceseoflascruces.org/document_page.php?num=

15. This act was reauthorized in 2005 and is now know as the Violence Against Women and Department of Justice Reauthorization Act of 2005, U.S. Government Printing Office, http://frwebgate.access.gpo.gov/cgi-bin/getdoc.cgi?dbname=109_cong_bills&docid=f:h3402enr.txt.pdf, December 14, 2009.

16. Interview no. 1, Sylvia, Las Cruces, New Mexico, November 10, 2000.

17. Interview no. 1, Mario, Las Cruces, New Mexico, December 5, 2000.

18. Interview no. 3, Marie, Valle de Vacas, New Mexico, November 2, 2000.

19. Marie, Community meeting, Valle de Vacas, June 6, 2000

20. Marie, Community meeting, Valle de Vacas, June 6, 2000

21. Interview no. 3, Estella, Recuerdos, New Mexico, October 30, 2000.

22. Mario is Estella's brother. They live in different colonias, but the colonia Estella lives in is one of the three Mario organizes.

23. Interview no. 1, Mario, Las Cruces, New Mexico, December 5, 2000.

24. If you remember, Sylvia was a colonia resident, an activist, and an organizer for more than ten years. She worked in the colonias and wanted to help colonia communities to improve and empower themselves.

25. Interview no. 1, Sylvia, Las Cruces, New Mexico, November 10, 2000.

26. Interview no. 1, Sylvia, Las Cruces, New Mexico, November 10, 2000.

27. Interview no. 1, John, Santa Fe, New Mexico, April 4, 2000.

28. Ibid.

29. Interview no. 2, Marie, Valle de Vacas, New Mexico, August 8, 2000.

30. Interview no. 1, Sylvia, Las Cruces, New Mexico, November 10, 2000.

31. Interview no. 3, Esperanza, Los Montes, New Mexico, October 31, 2000.

3. Empowerment and Politicization in the Colonias

1. Concha, Los Montes, New Mexico, 2000.

2. Main page UNIFEM, United Nations Development Fund for Women, http://www.unifem.org (accessed June 17, 2008).

3. "Politicization," like "empowerment," can be a very slippery term to use. My use of this term is in no way meant to imply that colonia activism on the individual scale is not political or that the colonias were not political before the COG arrived. Rather, the term "politicization" is a better fit for the goals of the COG. "Politicization," without empowerment's ties to the individual, can also be a better term to encompass the politics of marginalization, especially as marginalized groups relate to each other.

4. Interview no. 2, Juana, Recuerdos, New Mexico, September 27, 2000.

5. Interview no. 2, Estella, Recuerdos, New Mexico, September 19, 2000.

6. "Flushing" is the technical slang for connecting a community to a working waste-treatment plant. It literally means "flushing waste off of your property and out of your personal system."

7. Interview no. 2, Flora, Los Montes, New Mexico. August 14, 2000.

8. Interview no. 2, Juana, Recuerdos, New Mexico, September 27, 2000.

9. Interview no. 2, Esperanza, Los Montes, New Mexico, September 12, 2000.

10. Interview no. 3, Juana, Recuerdos, New Mexico, November 3, 2000.

11. Interview no. 2, Estella, Recuerdos, New Mexico, September 19, 2000.

12. Interview no. 3, Estella, Recuerdos, New Mexico, October 30, 2000.

13. Interview no. 2, Alicia, Valle de Vacas, New Mexico, September 5, 2000.

14. Elena, COG staff meeting, Las Cruces New Mexico, June 2005

15. I have to thank Michael Burawoy for an insightful and ultimately project-reshaping comment on a paper I presented almost immediately after I returned

from the field. In a room full of labor scholars, all generously funded by the University of California Institute for Labor and Employment, I was presenting on my study of women who often did not have paid employment. I was glowing with enthusiasm for the power of change these colonia activists employed daily, and Michael Burawoy smiled and said, "Where you see resistance I see consent, go back to Gramsci." I was not exactly crushed, but I certainly went home and delved deep into both my field notes and Gramsci's Prison Notebooks and was far from surprised to see Burawoy was absolutely correct.

16. Marlene, COG staff meeting, Las Cruces, New Mexico, June 2005

17. Ernie, personal communication, Las Cruces, New Mexico May 23, 2002.

4. The Place of NGOs in Daily Life

1. I use the terms "Mexican" and "American" here in the stereotypical form in which they are employed in the debates surrounding poverty and Mexican immigrants along the border (see Vila 2000).

2. Conversation COG grant writer, Las Cruces, New Mexico June 27, 2005.

Bibliography

Aparicio, A. 2006. *Dominican-Americans and the Politics of Empowerment*. Gainesville: University of Florida Press.

Bakker, I., ed. 1994. "Introduction: Engendering Macro-Economic Policy Reform in the Era of Global Restructuring and Adjustment." In *The Strategic Silence*. London: Zed.

Barry, A., T. Osborne, and N. Rose. 1996. *Foucault and Political Reason: Liberalism, Neoliberalism, and the Arts of Government*. Chicago: University of Chicago Press.

Bebbington, A. 1997. "Reencountering Development: Livelihood Transitions and Place Transformations in the Andes." *Annals of the Association of American Geographers* 90: 495–520.

Bourdieu, P. 1998. "The Essence of Neoliberalism." *Le Monde Diplomatique*, December.

Brown, W. 1995. *States of Injury: Power and Freedom in Late Modernity*. Princeton, N.J.: Princeton University Press.

Bryant, R. L. 2002. "Non-Governmental Organizations and Governmentality: 'Consuming' Biodiversity and Indigenous People in the Philippines." *Political Studies* 50: 268–92.

Burawoy, M., J. Blum, S. George, Z. Gille, M. Thayer, T. Gowen, L. Haney, M. Klawiter, S. Lopez, and S. Riain, ed. 2000. *Global Ethnography: Forces, Connections, and Imaginations in a Postmodern World*. Berkeley: University of California Press.

Burchell, G. 1996. "Liberal Government and Techniques of the Self." In *Foucault and Political Reason: Liberalism, Neoliberalism, and the Arts of Government*, ed. A. Barry, T. Osborne, and N. Rose. Chicago: University Press of Chicago.

Butler, J. 1999. *Gender Trouble: Feminism and the Subversion of Identity*. New York: Routledge.

Cable, S. 1992. "Women's Social Movement Involvement: The Role of Structural Availability in Recruitment and Participation Processes." *The Sociological Quarterly* 33: 35–50.

Calavita, K. 1996. "The New Politics of Immigration: 'Balanced-Budget Conservatism and the Symbolism of Proposition 187.'" *Social Problems* 43: 284–305.

Carew, J. 2001. "Minimum Standard Residential Subdivisions: Good Policy for Homestead Subdivisions?" Paper presented at Irregular Settlements and Self-Help Housing in the United States, Cambridge, Mass.

Chavez, L. R. 2001. *Covering Immigration: Popular Images and the Politics of Nation.* Berkeley: University of California Press.

Christopherson, S. 2002. "Changing Women's Status in a Global Economy." In *Geographies of Global Change,* 2nd ed., ed. R. J. Johnston, P. J. Taylor, and M. J. Watts, 191–205. Malden, Mass.: Blackwell.

Comaroff, J., and J. Comaroff. 2000. "Millennial Capitalism: First Thought on a Second Coming." *Public Culture* 20, no. 2: 291–343.

Conway, D. 2006. "Globalization from Below: Coordinating Global Resistance, Alternative Social Forums, Civil Society and Grassroots Networks." In *Globalization's Contradictions: Geographies of Discipline, Destruction and Transformation.,* ed. D. Conway, N. Heynen, 212–25. New York: Routledge.

Conway, D., and N. Heynen. 2006. "The Ascendancy of Neoliberalism and Emergence of Contemporary Globalization." In *Globalization's Contradictions: Geographies of Discipline, Destruction and Transformation.,* ed. D. Conway and N. Heynen, 17–34. New York: Routledge.

Craig, R. B. 1971. *The Bracero Program: Interest Groups and Foreign Policy.* Austin: University of Texas Press.

Cruikshank, B. 1999. *The Will to Empower: Democratic Citizens and Other Subjects.* Ithaca, N.Y.: Cornell University Press.

Cutter, S. L. 1995. "Race, Class, and Environmental Justice." *Progress in Human Geography* 19: 111–22.

Dean, M. 1999. *Governmentality: Power and Rule in Modern Society.* London: Sage Publications.

———. 2002. "Liberal Government and Authoritarianism." *Economy and Society* 31: 37–61.

Demmers, J. 2001. "Neoliberal Reforms and Populist Politics: The PRI in Mexico." In *Miraculous Metamorphoses: The Neoliberalization of Latin American Populism.,* ed. J. Demmers, A. E. Fernandez Joberto, and B. Hogenboom. New York: Zed Books.

Donelson, A. 2004. "The Role of NGOs and NGO Networks in Meeting the Needs of US Colonias." *Community Development Journal* 39, no. 4: 332–44.

Dolhinow, R. 2001. "When the Global Becomes Local: Colonia Development and the Social Reproduction of Labor on the Border." Paper presented at Irregular Settlements and Self-Help Housing Conference in the United States, Cambridge, Mass.

———. 2005. "Caught in the Middle: The State, NGOs, and the Limits to Grassroots Activism on the US–Mexico Border." *Antipode* 37, no. 3: 558–80.

Duggan, L. 2003. *The Twilight of Equality: Neoliberalism, Cultural Politics, and the Attack on Democracy.* Boston: Beacon Press.

Dunn, T. 1996. *The Militarization of the U.S.–Mexico Border 1978–1992: Low-Intensity Conflict Doctrine Comes Home.* Austin: University of Texas Press.

Edwards, M., and D. Hulme. 1996 "Too Close for Comfort? The Impact of Official Aid on Nongovernmental Organizations." *World Development* 24, no. 6: 961–73.

Elson, D., and R. Pearson. 1981. "Nimble Fingers Make Cheap Workers: An Analysis of Women's Employment in Third World Export Manufacturing." *Feminist Review* 7, (Spring): 87–107.

Esparza, A., and A. Donelson. 2008. *Colonias in Arizona and New Mexico: Border Poverty and Community Development Solutions.* Tucson: University of Arizona Press.

Freire, P. 2000. *Pedagogy of the Oppressed.* New York: Continuum.

Felch, J. 2002. "The Frozen Enchilada: Journalists Discuss a Shunned Mexico." *Center for Latin American Studies Newsletter.*

Feldman, R. M., S. Stall, and P. A. Wright. 1998. "The Community Needs to Be Built by Us: Women Organizing in Chicago Public Housing." In *Community Activism and Feminist Politics: Organizing across Race, Class, and Gender,* ed. N. A. Naples, 257–74. New York: Routledge.

Fisher, W. F. 1997. "Doing Good? The Politics of Antipolitics of NGO Practices." *Annual Review of Anthropology* 26: 439–64.

Foucault, M. 1970. *The Order of Things: An Archeology of the Human Sciences.* London: Tavistock.

———. 1990. "Two Lectures." In *Power/Knowledge: Selected Interviews and Other Writings 1972–1977,* ed. C. Gordon, 78–108. New York: Pantheon Books.

———. 1991. "Governmentality." In *The Foucault Effect: Studies in Governmentality,* ed. G. Burchell, C. Gordon, and P. Miller. Chicago: The University of Chicago Press.

———. 1994. "Governmentality." In *Michel Foucault: Power,* ed. J. D. Faubion, 201–22. New York: The New York Press.

Garcia y Griego, M. 1996. "The Importation of Mexican Contract Laborers to the United States, 1942–1964." In *Between Two Worlds: Mexican Immigrants in the United States,* ed. D. G. Gutierrez, 45–85. Wilmington, Del.: SR Books.

Gilmore, R. W. 1999. "You Have Dislodged a Boulder: Mothers and Prisoners in the Post Keynesian California Landscape." *Transforming Anthropology* 8: 12–38.

———. 2007. *Golden Gulag: Prisons, Surplus, Crisis, and Opposition in Globalizing California.* Berkeley: University of California Press.

Gjerde, J., ed. 1998. *Major Problems in American Immigration and Ethnic History.* New York: Houghton Mifflin Company.

Glick-Schiller, N., L. Basch, and C. Blanc-Szanton. 1992. "Transnationalism: A New Analytic Framework for Understanding Migration." *Annals of the New York Academy of Sciences* 645: 25–52.

Goldsmith, R. 1994. "Court Order Prevents Hatch from Evicting Farm Workers Until Civil Suit Heard." *Las Cruces Sun News.*

Gramsci, A. 1989. *Selections from the Prison Notebooks.* New York: International Publishers.

Gutman, M. C. 1996. *The Meanings of Macho: Being a Man in Mexico City.* Berkeley: University of California Press.

Hall, S. 1974. "Marx's Notes on Method: A 'Reading' of the '1857 Introduction.'" *Working Papers in Cultural Studies*: 132–70.

Hammack, D. 2002. "Nonprofit Organizations in American History." *American Behavioral Scientist* 45: 1638–74.

Hansen, T. B., and F. Stepputat. 2001. "Introduction: States of Imagination." In *States of Imagination: Ethnographic Explorations of the Postcolonial State*, ed. T. B. Hansen. Durham, N.C.: Duke University Press.

Harraway, D. 1988. "Situated Knowledges: The Science Question in Feminism and the Privilege of Partial Perspective." *Feminist Studies* 14: 575–99.

Hart, G. 2002. *Disabling Globalization: Places of Power in Post-Apartheid South Africa.* Berkeley: University of California Press.

———. 2004. "Development and Geography: Critical Ethnography." *Progress in Human Geography* 28, no. 1: 91–100.

———. 2006. "Denaturalizing Dispossession: Critical Ethnography in the Age of Resurgent Imperialism." *Antipode* 38, no. 3: 977–1004.

Harvey, D. 1990. *The Condition of Postmodernity.* Cambridge, Mass.: Blackwell.

———. 2005. *A Brief History of Neoliberalism.* New York: Oxford University Press.

Hill, S. 2003. "Metaphoric Enrichment and Material Poverty: The Making of 'Colonias.'" In *Ethnography on the Border*, ed. P. Vila, 141–65. Minneapolis: University of Minnesota Press.

Hintjens, H. 1999. "The Emperor's New Clothes: A Moral Tale for Development Experts?" *Development in Practice* 9, no. 3: 382–95.

Hondagneu-Sotelo, P. 2001. *Domestica: Immigrant Workers Cleaning and Caring in the Shadows of Affluence.* Berkeley: University of California Press.

Howell, J., and J. Pearce. 2001. *Civil Society and Development.* London: Lynne Rienner Publishers.

Hursh, D. 2007. "Assessing No Child Left Behind and the Rise of Neoliberal Education Policies." *American Education Research Journal* 44, no. 3: 493–518.

Hyatt, S. 2001. "From Citizen to Volunteer: Neoliberal Governance and the Erasure of Poverty." In *The New Poverty Studies: The Ethnography of Power, Politics, and Impoverished People in the United States*, ed. J. Goode, J. Maskovsky, 201–35. New York: New York University Press.

Jessop, B. 2002. "Liberalism, Neoliberalism and Urban Governance: A State-Theoretical Perspective." *Antipode* 34: 452–72.

Joseph, M. 2002. *Against the Romance of Community.* Minneapolis: University of Minnesota Press.

Kaplan, T. 1997. *Crazy for Democracy: Women in Grassroots Movements.* New York: Routledge.

Katz, C. 2001a. "On the Grounds of Globalization: A Topography for Feminist Political Engagement." *Signs: Journal of Women in Culture and Society* 26: 1309–234.

—————. 2001b. "Vagabond Capitalism and the Necessity of Social Reproduction." *Antipode* 33: 709–27.

—————. 2002. "Stuck in Place: Children and the Globalization of Social Reproduction." In *Geographies of Global Change: Remapping the World*, 2nd ed., ed. R. J. Johnston, P. J. Taylor, and M. J. Watts, 248–61. Malden, Mass.: Blackwell.

Keane, J. 1998. *Civil Society: Old Images, New Visions*. Stanford, Calif.: Stanford University Press.

Krauss, C. 1998. "Toxic Waste Protests and the Politicization of White, Working-Class Women." In *Community Activism and Feminist Politics: Organizing across Race, Class, and Gender*, ed. N. A. Naples, 129–50. New York: Routledge.

Larner, W. 2000. "Neo-liberalism: Policy, Ideology, Governmentality." *Studies in Political Economy* 63: 5–26.

—————. 2003. "Guest Editorial." *Environment and Planning D: Society and Space* 21: 509–12.

Larner, W., and D. Craig. 2005. "After Neoliberalism? Community Activism and Local Partnerships in Aotearoa, New Zealand." *Antipode* 37: 402–24.

Larson, J. E. 1995. "Free Market Deep in the Heart of Texas." *The Georgetown Law Journal* 84.

Lefebvre, H. 1996. *The Production of Space*. Cambridge, Mass.: Blackwell.

Li, T. 2007. *The Will to Empower: Governmentality, Development, and the Practice of Politics*. Durham, N.C.: Duke University Press.

Lorey, D. 1999. *The U.S.–Mexican Border in the Twentieth Century*. Wilmington, Del.: Scholarly Resources.

Manji, F. 1998. "The Depoliticization of Poverty." In *Development and Rights*, ed. D. Eade. Oxford: Oxfam.

Martin, D. 2004. "Nonprofit Foundations and Grassroots Organizing: Reshaping Urban Governance." *The Professional Geographer* 56, no. 3: 394–405.

Martin, P. 2005. "Comparative Topographies of Neoliberalism in Mexico." *Environment and Planning* 37, no. A: 203–20.

Massey, D. 1991. "The Political Place of Locality Studies." *Environment and Planning* A 23: 267–81.

—————. 1994. *Space, Place, and Gender*. Minneapolis: University of Minnesota Press.

Massey, D. S. 2002. *Beyond Smoke and Mirrors: Mexican Immigration in an Era of Economic Integration*. New York: Russell Sage Foundation.

McDowell, L. 1999. *Gender, Identity and Place: Understanding Feminist Geographies*. Minneapolis: University of Minnesota Press.

Mitchell, D. 2000. *Cultural Geography: A Critical Introduction*. Malden, Mass.: Blackwell.

Mitchell, K. 2004. *Crossing the Neoliberal Line: Pacific Rim Migration and the Metropolis*. Philadelphia: Temple University Press.

Mohan, G. 2002. "The Disappointments of Civil Society: The Politics of NGO Intervention in Northern Ghana." *Political Geography* 21: 125–54.

Moore, D. 1998. "Subaltern Struggles and the Politics of Place: Remapping Resistance in Zimbabwe's Eastern Highlands." *Cultural Anthropology* 13: 344–81.

Morris-Suzuki, T. 2000. "For and Against NGOs: The Politics of the Lived World." *New Left Review* 2.

Morton, A. 2003. "Structural Change and Neoliberalism in Mexico: 'Passive Revolution' in the Global Economy." *Third World Quarterly* 24, no. 2: 63–86.

Nagar, R. 2002. "Footloose Researchers, 'Traveling' Theories, and the Politics of Transnational Feminist Praxis." *Gender, Place and Culture* 9: 179–86.

Nagar, R., V. Lawson, L. McDowell, and S. Hanson. 2002. "Locating Globalization: Feminist (Re)readings of the Subjects and Spaces of Globalization." *Economic Geography* 78: 257–84.

Naples, N. A., 1998a. *Community Activism and Feminist Politics: Organizing across Race, Class, and Gender*. New York: Routledge.

———., ed. 1998b. *Grassroots Warriors: Activist Mothering, Community Work, and the War on Poverty*. New York: Routledge.

Nelson, M. 1995. "Colonias Proposals Opposed by County." *Las Cruces Sun News*.

Nevins, J. 2002. *Operation Gatekeeper: The Rise of the "Illegal Alien" and the Making of the U.S.–Mexico Boundary*. New York: Routledge.

Nichols, S. 2002. *Saints, Peaches, and Wine: Mexican Migrants and the Transformation of Los Haro, Zacatecas, and Napa*. Berkeley: University of California.

Ong, A. 1996. "Cultural Citizenship as Subject Making: Immigrants Negotiate Racial and Cultural Boundaries in the United States." *Current Anthropology* 37, no. 5: 737–62.

———. 2006. *Neoliberalism as Exception: Mutations in Citizenship and Sovereignty*. Durham, N.C.: Duke University Press.

Owen, G. 1988. *Las Cruces, New Mexico: 1849–1999 Multicultural Crossroads*. Las Cruces, NM: Red Sky Publishing.

Pardo, M. S. 1998a. *Mexican American Women Activists: Identity and Resistance in Two Los Angeles Communities*. Philadelphia: Temple University Press.

———. 1998b. "Creating Community: Mexican American Women in *Eastside* Los Angeles." In *Community Activism and Feminist Politics: Organizing across Race, Class, and Gender*, ed. N. A. Naples, 275–300. New York: Routledge.

Paster, E. 1993. *Colonias in Dona Ana County*, ed. County of Dona Ana, N.Mex: The Dona Ana County Planning Division.

Peck, J. 2004. "Geography and Public Policy: Constructions of Neoliberalism." *Progress Human Geography* 28, no. 3: 392–405.

Peck, J., and A. Tickell. 2002. "Neoliberalizing Space." *Antipode* 34, no. 2:380–404.

Peet, R. 2003. *Unholy Trinity: The IMF, World Bank, and WTO*. New York: Zed Books.

Petras, J., and H. Veltmeyer. 2001. *Globalization Unmasked: Imperialism in the 21st Century*. London: Zed Books.

Pred, A. 1995. "Out of Bounds and Undisciplined: Social Inquiry and the Current Moment of Danger." *Social Research* 62, no. 4: 1065–91.

———. 2000. *Even in Sweden: Racisms, Racialized Spaces, and the Popular Geographical Imagination*. Berkeley: University of California Press.

Pulido, L. 1996. *Environmental and Economic Justice: Two Chicano Struggles in the Southwest*. Tucson: The University of Arizona Press.

Rankin, K. 2004. *The Cultural Politics of Markets: Economic Liberalization and Social Change in Nepal*. Toronto: University of Toronto Press.

Reisler, M. 1996. "Always the Laborer, Never the Citizen: Anglo Perceptions of the Mexican Immigrant during the 1920s." In *Between Two Worlds: Mexican Immigrants in the United States*, ed. D. G. Gutierrez, 23–44. Wilmington, Del.: SR Books.

Riding, A. 1989. *Distant Neighbors: A Portrait of the Mexicans*. New York: Vintage Books.

Roelvink, G., and D. Craig. 2005. "The Man in the Partnering State: Regendering the Social Through Partnership." *Studies in Political Economy* 75: 103–26.

Rose, N. 1999. *Powers of Freedom: Reframing Political Thought*. Cambridge: Cambridge University Press.

Rowlands, J. 1997. *Questioning Empowerment: Working with Women in Honduras*. Oxford: Oxfam.

Rubin, J. W. 1997. *Decentering the Regime: Ethnicity, Radicalism, and Democracy in Juchitan, Mexico*. Durham, N.C.: Duke University Press.

Sassen, S. 1996. "U.S. Immigration Policy Toward Mexico in a Global Economy." In *Between Two Worlds: Mexican Immigrants in the United States*, ed. D. G. Gutierrez, 213–28. Wilmington, Del.: Scholarly Resources.

Sayer, A. 1991. "Beyond the Locality Debate: Deconstructing Geographer's Dualisms." *Environment and Planning A* 23: 283–308.

Skidmore, T. E., and P. H. Smith. 2001. *Modern Latin America*. 5th ed. New York: Oxford University Press.

Smith, N. 1993. "Contours of a Spatialized Politics: Homeless Vehicles and the Practice of Geographical Scale." *Social Text* 33: 54–81.

Stack, C. 1975. *All Our Kin: Strategies for Survival in a Black Community*. New York: Harper and Row.

Staudt, K. 1998. *Free Trade? Informal Economies at the U.S.–Mexico Border*. Philadelphia: Temple University Press.

Staudt, K., S. M. Rai, and J. L. Parpart. 2001. "Protesting World Trade Rules: Can We Talk about Empowerment?" *Signs: Journal of Women in Culture and Society* 26: 1251–57.

Stevenson, R. 2001. "Alternatives to Convention." Paper presented at the Irregular Settlement and Self-Help Housing in the United States Conference, Cambridge, Mass.

Tiano, S. 2006. "The Changing Composition of the Maquiladora Workforce Along the US–Mexico Border." In *Women and Change at the US–Mexican Border: Mobility, Labor and Activism*, ed. D. J. Mattingly and E. R. Hansen. Tucson: The University of Arizona Press.

Townsend, J., G. Porter, and E. Mawdsley. 2002. "The Role of the Transnational Community of Non-Governmental Organizations: Governance or Poverty Reduction?" *Journal of International Development* 14: 829–39.

Triantafillou, P., and M. R. Nielsen. 2001. "Policing Empowerment: The Making of Capable Subjects." *History of the Human Sciences* 14, no. 2: 63–86.

U.S. Environmental Protection Agency. 2000. "Protecting the Environment of the U.S.-Mexico Border Area." A briefing paper for the incoming U.S. administration. pp 1–19. http://www.scerp.org/transition.pdf

U.S. General Accounting Office (GAO). 1990. "Rural Development: Problems and Progress of Colonia Subdivisions Near Mexico Border." U.S. GAO Report RCED-91-37. Washington, D.C.

Vila, P. 2000. *Crossing Borders, Reinforcing Borders: Social Categories, Metaphors, and Narrative Identities on the U.S.–Mexico Frontier.* Austin: University of Texas Press.

Visweswaran, K. 1994. *Fictions of Feminist Ethnography.* Minneapolis: University of Minnesota Press.

Ward, P. 1999. *Colonias and Public Policy in Texas and Mexico: Urbanization by Stealth.* Austin: University of Texas Press.

———. 2000. *Residential Land Market Dynamics, Absentee Lot Owners and Densification Policies for Texas Colonias.* Austin: Lyndon Baines Johnson School of Public Policy.

———. 2001. "Dysfunctional Residential Land Markets: Colonias in Texas." *Land Lines.* January

Weedon, C. 1987. *Feminist Practice and Poststructuralist Theory.* New York: Basil Blackwell.

Williams, J. 2000. *Housing and Human Resources Needs Assessment.* Las Cruces: New Mexico State University.

Williams, R. 1977. *Marxism and Literature.* New York: Oxford University Press.

Wolch, J. R. 1989. "The Shadow State: Transformations in the Third Sector." In *The Power of Geography: The Shadow State: Transformations in the Voluntary Sector,* ed. J. R. Wolch and. J. Dear. Boston: Hyman.

Woods, C. 2002. "Life after Death." *Professional Geographer* 54, no. 1: 62–66.

Wright, M. 2006. *Disposable Women and Other Myths of Capitalism*. New York: Routledge.

Wright, R., and M. Ellis. 2000. "Race, Region and the Territorial Politics of Immigration in the US." *International Journal of Population Geography* 6: 197–211.

Wynia, G. 1990. *The Politics of Latin American Development*. New York: Cambridge University Press.

Index

Note: The italicized *f* following page numbers refers to figures.

REBECCA DOLHINOW is assistant professor of women's studies at the California State University at Fullerton.